SETS AND GROUPS

LIBRARY OF MATHEMATICS

edited by

WALTER LEDERMANN

D.Sc., Ph.D., F.R.S.Ed., Professor of
Mathematics, University of Sussex

SETS AND GROUPS

BY

J. A. GREEN

LONDON: Routledge & Kegan Paul Ltd

First published 1965
in Great Britain by
Routledge & Kegan Paul Ltd
Broadway House, 68–74 Carter Lane
London, E.C.4V 5EL

Reprinted 1966, 1967, 1971

ISBN 0 7100 4356 2

Printed in Great Britain
By Unwin Brothers Limited
The Gresham Press, Old Woking, Surrey, England
A Member of Staples Printing Group

Preface

MATHEMATICS is attractive and useful and limited because it attempts a precise expression of its ideas; people learn it in the hope of benefiting both from the ideas themselves and from the way they are expressed. But mathematics is always escaping from its own limits, and it produces new ideas and new language which represent a continual challenge to mathematical teaching.

The first three chapters of this book describe sets, relations and mappings; these are the basic ideas in terms of which most modern mathematics can be expressed. They are certainly 'abstract' ideas, but they are not difficult to grasp, because they come from notions which most people are used to from everyday experience. What is characteristic of modern mathematics is that it uses these few ideas over and over again, to build up theories which cover a great variety of situations. The rest of the book is an introduction to one such theory, namely the theory of groups.

The book is set out so that each new idea is described in its general form, and then illustrated by examples. These examples form a large part of the exposition, and require only a small amount of traditional mathematics to understand. There are also numerous exercises for the reader, at the end of each chapter.

It is a pleasure to thank my friends Drs. W. Ledermann, L. W. Morland and S. Swierczkowski, who each read and commented valuably on the manuscript.

J. A. GREEN

Contents

CONTENTS

CONTENTS

CHAPTER ONE

Sets

---

1. Sets

Elementary mathematics is concerned with counting and measurement, and these are described in terms of *numbers*. Modern 'abstract' mathematical disciplines start from the more fundamental idea of a *set*.

A set is simply a collection of things, which are called the *elements* or *members* of the set. We think of a set as a single object in its own right, and often denote sets by capital letters A, B, etc.

If A is a set, and if x is an element (or member) of A, we say x *belongs to* A. The symbol $\in$ is used for 'belongs to' in this sense, so that

$$x \in A$$

is a notation for the statement 'x is an element of A' or 'x belongs to A'. The negation of this, namely the statement 'x is not an element of A', is denoted $x \notin A$.

A set is determined by its elements, which means that a set A is fully described by describing all the elements of A. Sets A, B are defined to be *equal* (and we write $A = B$) if and only if they have the same elements, i.e. every element of A is an element of B, and every element of B is an element of A.

The notion of a set is very general, because there is virtually no restriction on the nature of the things which may be elements of a set. We give now some examples of sets.

SETS

Example 1. Let Z denote the set of all integers, i.e. the elements of Z are the integers (including zero and negative integers)

$$..., -2, -1, 0, 1, 2, ...$$

As examples of the $\in$ notation, we could write $0 \in Z$, $-4 \in Z$ and $\frac{1}{2} \notin Z$.

Example 2. Now let R denote the set of all real* numbers. Then for example $0, -4, \frac{1}{2}, \sqrt{2}, -\pi$ are some of the elements of R. A good way to visualize this set is to represent its elements by points on a straight line, as in co-ordinate geometry. A point on the line is chosen, arbitrarily, to represent the number 0, and then each element x of R is represented by the point at distance x from 0, putting positive x to the right, and negative x to the left of 0 in the usual way.

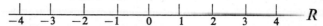

Throughout this book, the letters Z and R will be reserved for the two sets just described. See also the list of symbols on p. 81.

Example 3. Let P be the set of all people who are living at this moment. Then the reader of this page presumably belongs to P, but Julius Caesar does not, nor does the integer -4.

Example 4. Sometimes it is possible and convenient to display the elements of a set between brackets { }, for example $S = \{a, b, c\}$ means, S is the set whose elements are a, b and c. The order in which the elements are written does not matter, for example $\{b, a, c\}$ is the same set as S, since it has the same elements.

Example 5. A set A may have only one element, a, say, so that $A = \{a\}$. This 'one-element set' $\{a\}$ is logically different from a itself, because, for example, it is true that $a \in \{a\}$, but not that $a \in a$.

Finite and infinite sets. A set A which has only a finite number of elements is called a *finite set*, otherwise A is an *infinite set*. The number of elements in a finite set A is called the *order* of A and is denoted $|A|$.

Example 6. The sets Z and R are infinite. The set S of Example 4 is finite and $|S| = 3$. The set P of Example 3 is finite, but it is unlikely that its order will ever be known exactly!

A notation for sets. The following type of notation is often used to describe a set:

$$Z = \{x | x \text{ is an integer}\}.$$

---

* Ordinary numbers are often called *real* numbers, to distinguish them from complex numbers.

This is read 'Z is the set of all x such that x is an integer', which means the same as 'Z is the set of all integers'. In this notation, the symbol (such as x) on the left of the vertical line | stands for a typical element of the set, while on the right of the line is a statement about this typical element (such as the statement 'x is an integer') which serves to determine the set exactly.

Example 7. $Z(n) = \{x \in Z \mid 1 \leqslant x \leqslant n\}$ means, $Z(n)$ is the set of all x belonging to Z (i.e. of all *integers n*) such that $1 \leqslant x \leqslant n$. In other words, $Z(n)$ is the set whose elements are $1, 2, ..., n$. For example $Z(1) = \{1\}$, $Z(2) = \{1, 2\}, Z(3) = \{1, 2, 3\}$.

Example 8. $R^+ = \{x \in R \mid x > 0\}$ means, R^+ is the set of all *positive* real numbers.

2. Subsets

DEFINITION. Let A be any set. Then a set B is called a *subset* of A if every element of B is an element of A.

The notations $B \subseteq A$ and $A \supseteq B$ are both used to mean 'B is a subset of A'*. According to our definition, a set A is always a subset of itself, $A \subseteq A$. Any subset B of A which is not equal to A, is called a *proper* subset of A.

If $A \subseteq B$ and also $B \subseteq A$, then $A = B$, because then every element of A is an element of B and every element of B is an element of A.

Example 9. Z is a subset of R. The set B of all British people is a subset of the set P of all people. For any positive integer n, $Z(n)$ is a subset of Z. All these are proper subsets.

Example 10. Figure 1 is supposed to represent, diagrammatically, the relation $B \subseteq A$. A is represented by the set of all points inside the large circle, and B by the set of all points inside the small one. (Figures of this sort are sometimes called 'Venn diagrams').

Example 11. The notation $B \subseteq A$ is intended to recall the notation $b \leqslant a$ (meaning 'b is less than or equal to a') for numbers. But there is here an important difference between sets and numbers, namely that if A, B are two sets, it can easily happen that neither $A \subseteq B$ nor $B \subseteq A$ (for example, take $A = Z$ and $B = P$). But for any numbers a, b it is always true that $a \leqslant b$ or $b \leqslant a$.

---

* The symbol $\subset$ is often used instead of $\subseteq$.

The implication sign ⇒. The sign ⇒ is often used between two statements as an abbreviation for 'implies'. For example, if B is a subset of A we can write

$$x \in B \Rightarrow x \in A,$$

meaning 'x belongs to B implies x belongs to A', which is just another way of stating our definition 'Every element of B is an element of A'.

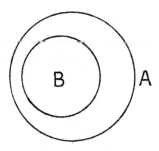

FIG. 1

If two statements are such that each implies the other, we write ⇔ between them. This sign can be read as 'implies and is implied by', or less clumsily 'if and only if'. Thus the condition that sets A, B should be equal may be written

$$x \in A \Leftrightarrow x \in B.$$

3. Intersection

DEFINITION. If A and B are sets, then the *intersection* $A \cap B$ of A and B is the set of all elements which belong to both A and B.

The notations introduced in the last two sections give us two other, but entirely equivalent, ways of putting this definition. First we might write

$$A \cap B = \{ x | x \in A \text{ and } x \in B \},$$

i.e. '$A \cap B$ is the set of all x such that $x \in A$ and $x \in B$'. Alternatively the condition that a given thing x belongs to $A \cap B$ can be expressed

$$x \in A \cap B \Leftrightarrow x \in A \text{ and } x \in B.$$

Example 12. If $S = \{a, b, c\}$ and $T = \{c, e, f, b\}$, then $S \cap T = \{b, c\}$. If B is the set of all British people, and C is the set of all blue-eyed people, then $B \cap C$ is the set of all blue-eyed, British people.

Example 13. Figure 2 shows $A \cap B$ as the shaded part which is in common to the circles representing A and B

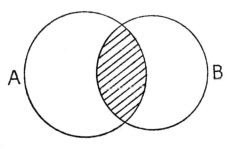

FIG. 2

For any sets A, B it is clear that $A \cap B = B \cap A$. Also $A \cap B$ is a subset of both A and B. In fact if Y is any set such that $Y \subseteq A$ and also $Y \subseteq B$, then $Y \subseteq A \cap B$ (because every element y of Y is an element of both A and B, hence $y \in A \cap B$). Thus $A \cap B$ is the 'largest' set which is a subset of both A and B.

Disjoint sets. The empty set. If sets A, B have no element in common, they are called *disjoint*. For example the sets Z and P of §1 are disjoint, because there is no integer which is also a person.

If A, B are disjoint, then $A \cap B$ is not really defined, because it has no elements. For this reason we introduce a conventional *empty set*, denoted $\emptyset$, to be thought of as a 'set with no elements'. Of course this is a set only by courtesy, but it is convenient to allow $\emptyset$ the status of a set. If A, B are disjoint, we write $A \cap B = \emptyset$.

5

The empty set is counted as a finite set, with order $|\varnothing| = 0$. Also $\varnothing$ must be regarded as a subset of any set A. For, by definition, $\varnothing \subseteq A$ means that every element of $\varnothing$ is an element of A. Since $\varnothing$ has no elements, this last statement is certainly true!

4. Union

DEFINITION. If A and B are sets, then the *union* $A \cup B$ of A and B is the set of all elements of A, together with all elements of B. Alternative expressions of this definition are

$$A \cup B = \{x \,|\, x \in A \text{ or } x \in B\}, \text{ or}$$

$$x \in A \cup B \Leftrightarrow x \in A \text{ or } x \in B,$$

where in both statements, the word 'or' is used in a sense which includes 'and', i.e. $A \cup B$ is the set of all elements which belong to A or to B, *including* those which belong to both A and B. Thus $A \cap B$ is always a subset of $A \cup B$ (see Figure 2, where $A \cup B$ is the whole region bounded by the two circles, including the shaded part).

Example 14. Let S, T, B, C be as in Example 12. Then $S \cup T = \{a, b, c, e, f\}$, and $B \cup C$ is the set of people who are either British, or blue-eyed, or both.

Example 15. For any sets A, B it is clear that $A \cup B = B \cup A$. Also both A and B are subsets of $A \cup B$. In fact if X is any set such that $A \subseteq X$ and $B \subseteq X$, then $A \cup B \subseteq X$ (because every element x of $A \cup B$ is an element of A or else an element of B, and in either case $x \in X$). Thus $A \cup B$ is the 'smallest' set in which both A and B are subsets.

Example 16. Let A, B be finite sets. Then

$$|A \cup B| = |A| + |B| - |A \cap B|.$$

For we can count the number of elements in $A \cup B$, by first counting the $|A|$ elements of A, and then the $|B|$ elements of B; but then we have counted twice each of the elements of $A \cap B$, so we must subtract $|A \cap B|$ from $|A| + |B|$ to get the total.

5. The algebra of sets

The operations $A \cap B$, $A \cup B$ are in some ways like the operations ab, $a + b$ with numbers. In the 'algebra of sets'* we can prove identities, as in ordinary algebra. Two simple examples of such *set-theoretic identities* are $A \cap B = B \cap A$, and $A \cup B = B \cup A$, both of which hold for all sets A, B. These resemble the algebraic identities $ab = ba$ and $a + b = b + a$, which hold for all numbers a, b. The identity proved in Example 17, below, resembles the algebraic identity $a(b + c) = ab + ac$.

This analogy between sets and numbers does not go very far. For example $A \cap A = A$ is an identity for sets, but its algebraic counterpart $aa = a$ is not an identity, since it does not hold for all numbers a.

There is an interesting *duality principle* for sets, according to which any identity involving the operations $\cap$, $\cup$ (and no others) remains valid if the symbols $\cap$, $\cup$ are interchanged throughout (see Example 18 for an illustration of this). But there is no corresponding 'duality' between ordinary multiplication and addition. For example, if we interchange addition and multiplication in the identity $a(b + c) = (ab) + (ac)$ (we have put in some extra brackets to make this easier to do) we get $a + (bc) = (a+b)(a+c)$, and this is not a correct algebraic identity.

Example 17. To prove that $A \cap (B \cup C) = (A \cap B) \cup (A \cap C)$, for all sets A, B, C. *Proof.* The reader should check each step below, remembering that $\Leftrightarrow$ means 'if and only if': $x \in A \cap (B \cup C) \Leftrightarrow x \in A$ and $x \in B \cup C \Leftrightarrow x \in A$ and $(x \in B$ or $x \in C) \Leftrightarrow (x \in A$ and $x \in B)$ or $(x \in A$ and $x \in C) \Leftrightarrow x \in A \cap B$ or $x \in A \cap C \Leftrightarrow x \in (A \cap B) \cup (A \cap C)$. This proves that every element x of $A \cap (B \cup C)$ is an element of $(A \cap B) \cup (A \cap C)$, and conversely. Hence the two sets are equal.

Example 18. The 'dual' of the last identity is $A \cup (B \cap C) = (A \cup B) \cap (A \cup C)$. To prove this, we can 'dualize' the proof just given, by interchanging the symbols $\cap$, $\cup$ and also the words 'and', 'or', throughout. The reader will find that this gives automatically a proof of this new identity.

---

* Invented by G. Boole (1815–1864).

6. Difference and complement

DEFINITION. If A and B are sets, then the *difference set $A - B$* is the set of those elements of A which do not belong to B. Alternative expressions of this definition are

$$A - B = \{x \mid x \in A \text{ and } x \notin B\}, \text{ or}$$

$$x \in A - B \Leftrightarrow x \in A \text{ and } x \notin B.$$

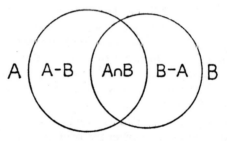

FIG. 3

Complement. If B is a subset of A, then $A - B$ is sometimes called the *complement* of B in A. Clearly B is the complement of $A - B$ in A (see Figure 4).

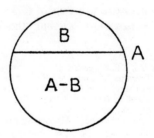

FIG. 4

Example 19. Figure 3 shows the relative situations of the three sets $A - B$, $A \cap B$ and $B - A$. If $A \subseteq B$ it is clear that $A - B = \varnothing$, in particular $A - A = \varnothing$.

Example 20. For any sets A and B, $A - B = A - (A \cap B)$. If A is finite and B is a subset of A, then $|A - B| = |A| - |B|$. This formula shows why the sign $-$ is used for difference of sets, but it should be remembered that $A - B$ is defined, even when B is not a subset of A.

Example 21. Sets A_1, A_2, A_3 are represented in Figure 5 by the interiors of the circles shown. $A_1 - A_2$ is the region shaded horizontally, and $A_1 - A_3$ is shaded vertically. Their intersection is B_1, and the diagram shows that this is also $A_1 - (A_2 \cup A_3)$. This suggests (but does not prove!) the identity

$$(A_1 - A_2) \cap (A_1 - A_3) = A_1 - (A_2 \cup A_3).$$

We prove this formally, for any sets A_1, A_2, A_3, as follows: $x \in (A_1 - A_2) \cap (A_1 - A_3) \Leftrightarrow x \in A_1 - A_2$ and $x \in A_1 - A_3 \Leftrightarrow (x \in A_1$ and $x \notin A_2)$ and $(x \in A_1$ and $x \notin A_3) \Leftrightarrow x \in A_1$ and x belongs to neither A_2 nor $A_3 \Leftrightarrow x \in A_1$ and $x \notin A_2 \cup A_3 \Leftrightarrow x \in A_1 - (A_2 \cup A_3)$.

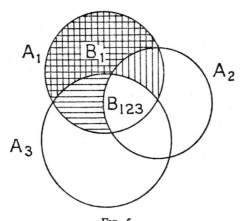

Fig. 5

7. Pairs. Product of sets

In ordinary plane co-ordinate geometry, a point is described by two co-ordinates x, y, that is, the point is described by a *pair* (x, y) of numbers.

The pair (x, y) is not the same as the *set* $\{x, y\}$ which has x and y as its elements. In a pair, the order of the terms is essential (in fact the term *ordered pair* is often used to

9

emphasize this). For example, (1, 2) and (2, 1) represent different points in the plane (see Figure 6), but the sets $\{1, 2\}$ and $\{2, 1\}$ are the same, because they have the same elements. Again, we may have a pair (x, y) with $x = y$, for example $(0, 0)$ or $(1, 1)$.

We have mentioned co-ordinate geometry as an illustration, but now we shall formulate the idea of a pair in general terms. If x and y are any things (they need not be numbers, and they need not be different) let us use the symbol (x, y) to denote the pair consisting of x and y in that order. Pairs (x, y) and (x', y') are considered *equal* if and only if $x = x'$ and $y = y'$. x is called the first, and y the second, *component* of (x, y).

DEFINITION. If A and B are sets, then the *product* (sometimes called *Cartesian product*) $A \times B$ of A and B is the set of all pairs (x, y) such that $x \in A$ and $y \in B$.

Example 22. Let $A = \{1, 2, 3\}$ and $B = \{a, b\}$. Then $A \times B$ has six elements, viz. $(1, a)$, $(2, a)$, $(3, a)$, $(1, b)$, $(2, b)$, $(3, b)$. $B \times A$ also has six elements, obtained by reversing these, viz. $(a, 1)$, $(a, 2)$, $(a, 3)$, $(b, 1)$, $(b, 2)$, $(b, 3)$. If A, B are any finite sets, then $|A \times B| = |A|\,|B|$, because there are $|A|$ ways of choosing the first component of an element (x, y) of $A \times B$, and, for each such choice, $|B|$ ways of choosing the second component. This is the reason for using the sign $\times$ for product of sets.

Example 23. If A is any set, then $A \times A$ is the set of all pairs (x, y), where both x and y belong to A. For example, if R is the set of all real numbers, then $R \times R$ is exactly the set of all the pairs (x, y) which represent points in plane co-ordinate geometry. Plane curves and other figures can be regarded as subsets of $R \times R$. For example the set $E = \{(x, y)|y = x^2\}$ is the subset of $R \times R$ represented on Figure 6 by the curve with equation $y = x^2$, because this is just the set of those points (x, y) such that this equation holds. The set shaded represents the subset $F = \{(x, y)|y \leqslant x^2\}$ of $R \times R$.

8. Sets of sets

We shall find, later in this book, that we often have to deal with a *set of sets*, that is, with a set whose elements are themselves sets. Sometimes we shall use script capitals, such as $\mathscr{P}$, $\mathscr{S}$, etc. to denote such sets of sets. The only new feature is that we must be careful how we use the sign $\in$. If $\mathscr{S}$ is a set of sets, and if $A \in \mathscr{S}$, then A is itself a set. But an element, say x, of A is *not* usually an element of $\mathscr{S}$. Thus $x \in A$ and $A \in \mathscr{S}$, but in general $x \notin \mathscr{S}$.

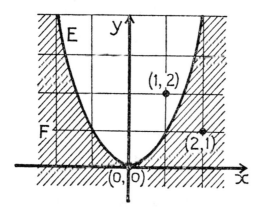

FIG. 6

Example 24. Let $A = \{1, 2, 3, 4\}$, $B = \{a, 3, 4\}$, $C = \{2, a\}$. Then $\mathscr{S} = \{A, B, C\}$ is a set of sets. It has three elements A, B, C. The elements $1, 2, 3, 4, a$ of these sets are not elements of $\mathscr{S}$.

Example 25. If X is a given set, we shall always denote by $\mathscr{B}(X)$ the set of *all subsets* of X. In particular $\varnothing$ and X itself are elements of $\mathscr{B}(X)$. As an example take $X = \{x, y, z\}$. Then $\mathscr{B}(X)$ has 8 elements $\varnothing$, $\{x\}$, $\{y\}$, $\{z\}$, $\{x, y\}$, $\{x, z\}$, $\{y, z\}$, $\{x, y, z\}$. Notice that $\{x\} \in \mathscr{B}(X)$ but $x \notin \mathscr{B}(X)$.

Intersection and union. Let $\mathscr{S}$ be any set of sets. We define the intersection $\bigcap \mathscr{S}$ to be the set of all objects x which belong to all of the sets A in $\mathscr{S}$, i.e.

$$\bigcap \mathscr{S} = \{x \mid x \in A \text{ for all } A \in \mathscr{S}\}.$$

Similarly the union $\bigcup \mathscr{S}$ is defined to be the set of all objects x which belong to at least one set A in $\mathscr{S}$, i.e.

$$\bigcup \mathscr{S} = \{x \mid x \in A \text{ for at least one } A \in \mathscr{S}\}.$$

SETS

If $\mathscr{S}$ has only two members, $\mathscr{S} = \{A, B\}$, then these definitions reduce to our earlier definitions for $A \cap B$ and $A \cup B$ respectively, i.e. $\bigcap \{A, B\} = A \cap B$ and $\bigcup \{A, B\} = A \cup B$. If $\mathscr{S}$ is a finite set of sets, say $\mathscr{S} = \{A_1, ..., A_n\}$, we often write

$$\bigcap \mathscr{S} = A_1 \cap ... \cap A_n, \text{ and}$$

$$\bigcup \mathscr{S} = A_1 \cup ... \cup A_n.$$

We shall not, in fact, use the notations $\bigcap$ and $\bigcup$ very much in this book.

Example 26. Take $\mathscr{S} = \{A, B, C\}$ as in Example 24. Then $\bigcap \mathscr{S} = A \cap B \cap C = \emptyset$ and $\bigcup \mathscr{S} = A \cup B \cup C = \{1, 2, 3, 4, a\}$. Notice that $A \cap B \cap C$ is empty, although none of $A \cap B$, $B \cap C$, $A \cap C$ is.

Example 27. In Figure 5, $A_1 \cap A_2 \cap A_3$ is represented by the region B_{123}, and $A_1 \cup A_2 \cup A_3$ by the whole figure. Notice

$$(A_1 \cap A_2) \cap A_3 = A_1 \cap A_2 \cap A_3 = A_1 \cap (A_2 \cap A_3).$$

and the dual identities for unions.

Example 28. As example of an infinite set of sets take $\mathscr{S} = \{Z(n) | n = 1, 2, 3, ... \}$, i.e. the *elements* of $\mathscr{S}$ are the sets $Z(1) = \{1\}$, $Z(2) = \{1, 2\}$, $Z(3) = \{1, 2, 3\}$, etc. (Notice that $\mathscr{S}$ is a *subset* of $\mathscr{B}(Z)$.) It is easy to see that $\bigcap \mathscr{S} = \{1\}$ and $\bigcup \mathscr{S}$ is the set of all positive integers.

Russell's paradox*. Serious difficulties occur if we allow the notion of set to be too general. For example it is undesirable to talk of 'the set U of all sets'. If such a set exists, it must, being a set, be a member of itself, $U \in U$. This is not in itself disastrous, but now consider Russell's famous set V of *all sets which are not members of themselves*. If V is not a member of itself, then by that fact it qualifies as member of V, i.e. it *is* then a member of itself. And the situation is no better if we yield to this reasoning and allow that V *is* a member of itself. For then V itself does not qualify as member of V, which only admits as members those sets which are not members of themselves, so we find that we have again proved the opposite of what we assumed. This logical impasse can be avoided by restricting the notion of set, so that 'very large' collections such as U or V or the 'collection of all things' are not counted as sets. However this is done at some cost in simplicity, and in this book we shall do no more than keep to sets which appear to be harmless, and hope that paradoxes will not appear.

---

* Invented by Bertrand Russell in 1901.

SETS

EXERCISES FOR CHAPTER ONE

1. Describe in words the sets $A = \{x \in Z \,|\, 2 \leqslant x\}$, $B = \{x \in Z \,|\, x \leqslant 5\}$. Show that $A \cap B$ is finite and that $A \cup B = Z$.

2. If A and B are any sets, prove that $A \cap B = B$ if and only if $A \supseteq B$.

3. Prove the identities $(A \cap B) \cup A = A$ and $(A \cup B) \cap A = A$.

4. Let A_1, A_2, A_3 be finite sets, and write $a_i = |A_i|$, $a_{ij} = |A_i \cap A_j|$ and $a_{123} = |A_1 \cap A_2 \cap A_3|$. Prove that $|A_1 \cup A_2 \cup A_3| = a_1 + a_2 + a_3 - a_{12} - a_{13} - a_{23} + a_{123}$.

5. With the notation of Exercise 4, suppose that $a_1 = 10$, $a_2 = 15$, $a_3 = 20$, $a_{12} = 8$ and $a_{23} = 9$. Prove that the only values which $|A_1 \cup A_2 \cup A_3|$ could have are 26, 27 or 28.

6. Prove the identity $(A_1 - A_2) \cup (A_1 - A_3) = A_1 - (A_2 \cap A_3)$.

7. For any sets A, B define the 'symmetric sum' $A \oplus B$ to be the set $A \cup B - (A \cap B)$. Draw a diagram to represent $A \oplus B$. Prove the identities (i) $A \oplus B = B \oplus A$; (ii) $A \oplus A = \varnothing$; (iii) $(A \oplus B) \oplus C = A \oplus (B \oplus C)$; (iv) $(A \oplus B) \cap C = (A \cap C) \oplus (B \cap C)$.

8. Prove the identities (i) $A \times (B \cap C) = (A \times B) \cap (A \times C)$ and (ii) $A \times (B \cup C) = (A \times B) \cup (A \times C)$. Show that the sets $(A \times B) \times C$ and $A \times (B \times C)$ are never the same, and that $A \times B = B \times A$ if and only if $A = B$.

9. Mark on a co-ordinate plane the regions which represent the following subsets of $R \times R$ (see Example 23, p. 10): (i) $\{(x, y) \,|\, x = 0\}$, (ii) $\{(x, y) \,|\, x > y\}$, (iii) $\{(x, y) \,|\, x^2 + y^2 = 1\}$, (iv) $\{(x, y) \,|\, x^2 + y^2 < 1\}$, (v) $\{(x, y) \,|\, 0 \leqslant x \leqslant 1, 0 \leqslant y \leqslant 1\}$.

10. Prove that $(A \times B) \cap (B \times A) = (A \cap B) \times (A \cap B)$, for any sets A, B. Is the same true with $\cup$ replaced by $\cap$?

11. Prove the identity $(A \cup B) \times (C \cup D) = (A \times C) \cup (A \times D) \cup (B \times C) \cup (B \times D)$. (This compares with the algebraic identity $(a + b)(c + d) = ac + ad + bc + bd$.)

12. For each positive integer n, let $Q_n = \{x \in R \,|\, 0 \leqslant x < (n + 1)/n\}$. Let $\mathscr{S}$ be the set of all these sets Q_n, for $n = 1, 2, \ldots$. Show that $\cap \mathscr{S} = \{x \in R \,|\, 0 \leqslant x \leqslant 1\}$. Find also $\cup \mathscr{S}$.

13. Prove that the number of subsets (including $\varnothing$) of a finite set of order n is 2^n. (Use induction on n. See also Example 35, p. 17.)

CHAPTER TWO

Equivalence Relations

---

1. Relations on a set

In many sets there occur relations which hold between certain pairs of elements. A relation is usually described by a *statement* which involves an arbitrary pair of elements of the set. For example the statement 'x is the mother of y' describes a relation on the set P of all living people; for certain pairs (x, y) of people the statement is true, and for all other pairs it is false. Similarly '$x > y$' (meaning 'x is greater than y') describes a relation on the set R of all real numbers.

As a general notation we shall often use $\sim$ for a relation on a set A.

DEFINITION. $\sim$ is a *relation* on the set A*, if, for each pair (x, y) of elements of A, the statement '$x \sim y$' has a meaning, i.e. it is either true or false for that particular pair. We write simply $x \sim y$, to mean that '$x \sim y$' is *true*, for a given pair (x, y).

For example, taking $A = P$, we could use '$x \sim y$' as an abbreviation for 'x is the mother of y', and then $\sim$ is a relation on P. Similarly $>$ is a relation on R, because for every pair (x, y) of real numbers, the statement '$x > y$' has a meaning, i.e. it is either true or false for that particular pair. When we write simply $x > y$, with no quotation marks, then of course we mean that the statement '$x > y$' is *true*.

A very familiar relation, applicable to any set A, is the relation of *equality*, denoted by $=$, so that '$x = y$' means, as usual, 'x is equal to y'.

---

* Often called a *binary relation*, because it relates pairs of elements of A.

14

2. Equivalence relations

Now we shall describe some special kinds of relation. Let A be any set, and let $\sim$ be a relation on A.

DEFINITION 1. $\sim$ is a *reflexive* relation if it satisfies the condition

E1. If x is any element of A, then $x \sim x$.

DEFINITION 2. $\sim$ is a *symmetric* relation if its satisfies

E2. If x, y are any elements of A such that $x \sim y$, then also $y \sim x$.

DEFINITION 3. $\sim$ is a *transitive* relation if its satisfies

E3. If x, y, z are any elements of A such that $x \sim y$ and $y \sim z$, then also $x \sim z$.

DEFINITION 4. $\sim$ is an *equivalence relation** if it is reflexive, symmetric and transitive, i.e. if it satisfies all three conditions **E1, E2, E3.**

Example 29. The relation of equality is the simplest equivalence relation. If we take $x \sim y$ to mean $x = y$, clearly **E1, E2, E3** are satisfied.

Example 30. Define a relation $\sim$ on the set P of all people, by taking '$x \sim y$' to mean 'x and y have the same age' (meaning, to be precise, that their ages on their last birthdays were the same). This is an equivalence relation on P, as we see at once by checking that **E1, E2, E3** are satisfied.

Example 31. If we define $\sim$ to be the relation 'is the mother of' on the set P, we find that this is neither reflexive, symmetric nor transitive. Therefore it is not an equivalence relation on P.

Example 32. The relation $>$ on R is *transitive*, for if x, y, z are any numbers such that $x > y$ and $y > z$, clearly also $x > z$. But it is not reflexive, since '$x > x$' is never true, and it is not symmetric (take for example $x = 2$, $y = 1$ to show that **E2** is not satisfied). So $>$ is not an equivalence relation.

3. Partitions

In the next section we shall prove that any equivalence relation on a set A determines a *partition* of A. So we must interrupt our discussion of relations, to define this new idea.

DEFINITION. A *partition* of a set A is a set $\mathscr{P}$ of non-empty subsets of A, such that each element x of A belongs to one and only one member of $\mathscr{P}$.

---

* The symbols $\simeq$, $\cong$, $\equiv$ are often used to denote equivalence relations.

Figure 7 illustrates the way a partition 'divides up' the set A. In this case $\mathscr{P}$ is the set with five elements X_1, X_2, X_3, X_4, X_5, which are subsets of A, such that each element of A lies in one and only one of them.

Example 33. Let Z^+ be the set of all positive integers, Z^- the set of all negative integers, and Z^0 the one-element set $\{0\}$. Then $\mathscr{P} = \{Z^+, Z^-, Z^0\}$ is a partition of the set Z of all integers, because every integer z belongs to one and only one of Z^+, Z^- and Z^0.

Example 34. We get a partition of a set A as soon as we *classify* the elements of A in such a way that each element x of A falls into one and only one 'class', i.e. subset. For example, we might classify people according to their age on their last birthday, and let X_n denote the set of all people of age n; then the set of all these sets X_n is a partition of the set P of all people.

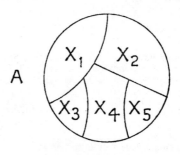

FIG. 7

Partitions of a finite set. Suppose that $\mathscr{P}$ is a partition of a finite set A into n subsets, $\mathscr{P} = \{X_1, \ldots, X_n\}$, say. By counting up all the elements in these sets $X_1, \ldots, X_n$ we must count each element of A, and of course we have counted no element twice, because we know that no element x of A belongs to two of the sets X_1. This gives the simple formula

$$|A| = |X_1| + \ldots + |X_n|,$$

which is often very useful.

16

Example 35. Let $B = \mathscr{B}(X)$ be the set of all subsets of a set X which has n elements. For each integer r in the range $0 \leqslant r \leqslant n$, let Y_r be the set of all subsets of X which have order r. For example, Y_0 contains only the empty set $\varnothing$, Y_1 is the set of all the one-element subsets of X, and so on. Now $\{Y_0, Y_1, ..., Y_n\}$ is a partition of B, because every element of B belongs to exactly one of these sets, hence $|B| = |Y_0| + |Y_1| + ... + |Y_n|$. From elementary algebra, $|Y_r|$, the number of ways of choosing a set of r elements from the n elements of X, is the 'binomial coefficient'

$$\binom{n}{r} = \frac{n!}{r!(n-r)!}$$

(often written ^nC_r), i.e. it is the coefficient of t^r in the expansion of $(1 + t)^n$ in powers of t. Thus $|Y_0| + |Y_1|t + |Y_2|t^2 + ... + |Y_n|t^n = (1 + t)^n$, and putting $t = 1$ we find

$$|B| = |\mathscr{B}(X)| = |Y_0| + |Y_1| + ... + |Y_n| = 2^n.$$

See also Example 57, p. 33.

4. Equivalence classes

DEFINITION. Let A be any set, and $\sim$ any equivalence relation on A. For each fixed element x of A, the *equivalence class* E_x of x (with respect to the relation $\sim$) is the set of all elements y of A such that $x \sim y$, i.e.

$$E_x = \{y \in A \,|\, x \sim y\}.$$

Thus E_x is a subset of A. Since $x \sim x$ by **E1**, we see that x itself belongs to E_x.

Example 36. Let $\sim$ be the equivalence relation on P defined in Example 30, p. 15. If x is a given person, then E_x (with respect to this relation) is the set of all people y who have the same age as x; if the age of x is n years, then E_x is the set X_n of Example 34. Notice that E_y also is the same set, for any other person y of the same age, in fact, $E_x = E_y$ if and only if $x \sim y$. The equivalence classes E_x, for all $x \in P$, are the sets X_n of Example 34, and these form a partition of P. The reader may find this example useful in understanding the theorem which follows.

FUNDAMENTAL THEOREM ON EQUIVALENCE RELATIONS

If $\sim$ is an equivalence relation on a set A, then the set $\mathscr{P} = \{E_x \,|\, x \in A\}$ of all equivalence classes with respect to $\sim$ is a partition of A. Moreover if x, y are any elements of A, then $E_x = E_y$ if and only if $x \sim y$.

We shall prove this in three steps, in the first two of which are proved two 'lemmas', i.e. auxiliary facts needed in the final step.

(i) LEMMA. *If $x \sim y$, then $E_x = E_y$.*

Proof. Suppose that $x \sim y$, and that z is any element of E_y, which means that $y \sim z$. Now we have $x \sim y$ and $y \sim z$, therefore $x \sim z$ by E3. Hence $z \in E_x$. Thus every element of E_y is an element of E_x. But the symmetric property E2 shows that also $y \sim x$, and the same argument shows that every element of E_x is an element of E_y. Therefore $E_x = E_y$.

(ii) LEMMA. *Any two equivalence classes which are not equal, are disjoint, i.e. if $E_x \neq E_y$ then $E_x \cap E_y = \varnothing$.*

Proof. This is a proof by contradiction; we start by supposing that E_x, E_y are not disjoint. Then there is an element z of A which belongs to both, i.e. $x \sim z$ and also $y \sim z$. Applying lemma (i), this means $E_x = E_z$ and also $E_y = E_z$, hence $E_x = E_y$ which contradicts our assumption. Therefore our supposition that E_x, E_y are *not* disjoint must be false, and this proves the lemma.

(iii) *Proof of the theorem.* Each element x of A belongs to at least one equivalence class, namely to E_x. And if x belonged also to an equivalence class E_y which is not equal to E_x, this would contradict lemma (ii), for E_x and E_y are not disjoint if they both contain x. This proves that $\mathscr{P}$ is a partition of A, because each element of A belongs to one and only one member of $\mathscr{P}$.

Now we have to prove that $E_x = E_y$ if and only if $x \sim y$. Lemma (i) already shows that $E_x = E_y$ if $x \sim y$. Suppose conversely that $x \sim y$. Then $y \in E_x$, but also $y \in E_y$, hence E_x, E_y are not disjoint, and lemma (ii) shows that $E_x = E_y$. This completes the proof of the theorem.

Example 37. Suppose we start with any partition $\mathscr{Q}$ of a set A. Then we may define a relation $\sim$ on A by the rule, '$x \sim y$' shall mean 'x and y belong to the same member of $\mathscr{Q}$'. It is easy to verify that this is an equivalence relation on A, and that the equivalence classes with respect to

18

$\sim$ are precisely the sets which are members of $\mathcal{Q}$. This provides a converse to the theorem above, and shows that *equivalence relations* and *partitions*, of a given set A, are practically interchangeable.

5. Congruence of integers

For the rest of this chapter we shall discuss in more detail a particular example of an equivalence relation, which is very useful in the theory of numbers (i.e. of integers), and has given rise to many important ideas in algebra

Let m be a positive integer; we shall keep m fixed for the rest of the chapter. If x, y are any integers we make the DEFINITION. x is congruent to y modulo m, if $x - y$ is a multiple of m, i.e. if there exists an integer k such that $y = x + km$. We denote this*

$$x \equiv y \bmod m.$$

It is easy to verify that, for fixed m, this relation of 'congruence mod m' is an *equivalence relation on the set Z of all integers*.

E1 holds, because for any $x \in Z$, $x - x = 0$ is a multiple of m, hence $x \equiv x \bmod m$. **E2**. Suppose $x \equiv y \bmod m$, i.e. $x - y$ is a multiple of m. Then the same is true of $y - x = -(x - y)$, so $y \equiv x \bmod m$. **E3**. Suppose $x \equiv y$ and $y \equiv z \bmod m$, i.e. $x - y$ and $y - z$ are multiples of m. Then the same is true of $x - z = (x - y) + (y - z)$, and so $x \equiv z \bmod m$.

Example 38. $13 \equiv 3 \bmod 5$, $-10 \equiv 8 \bmod 6$, $-3 \equiv -5 \bmod 2$. Any two integers x, y are congruent mod 1. To say that $x \equiv 0 \bmod m$ is the same as saying x is a multiple of m. If $x \equiv y \bmod m$, and if n is a factor (divisor) of m, then $x \equiv y \bmod n$.

Congruence classes (*residue classes*). Let x be a given integer. The equivalence class E_x with respect to the relation of congruence mod m, is called the *congruence class*, or *residue class of x* mod m. E_x is the set of all integers y such that $x \equiv y \bmod m$, i.e. $E_x = \{x + km | k \in Z\} =$

$$\{..., -2m + x, -m + x, x, x + m, x + 2m, ...\}.$$

---

* 'Modulo' means 'with modulus (or measure)'. The idea of congruence of integers was introduced by Gauss (1777–1855).

We use 'mod m' as an abbreviation for 'modulo m'.

Let r be the smallest integer r in E_x which is $\geqslant 0$; r is called the *residue of x* mod m. r must be one of the integers $0, 1, ..., m-1$, because if $r \geqslant m$, then $r-m$ is also in E_x and is $\geqslant 0$, but is smaller than r. Also $E_x = E_r$, by the fundamental theorem on p. 17, because $x \equiv r$ mod m. Thus each class E_m is equal to one of the m classes $E_0, E_1, ..., E_{m-1}$; moreover these are distinct sets, because no two of $0, 1, ... m-1$ are congruent mod m. To summarize, *there are exactly m different congruence classes mod m, namely $E_0, E_1, ..., E_{m-1}$, and these form a partition of Z. The class E_r consists of all integers which have a given residue r mod m.*

Example 39. If x is positive, then the residue r is the same as the *remainder on division by m*. For when we divide x by m we subtract as large a multiple of m as we can, say qm, from x, without making the difference $x - qm$ negative. But $x - qm$ belongs to E_x, for all $q \in Z$, so the remainder $r = x - qm$ is the smallest element of E_x which is $\geqslant 0$. For example, on dividing 29 by 8 we get 'quotient' $q = 3$, so $5 = 29 - 3 \cdot 8$ is the residue of 29 mod 8.

Example 40. Taking $m = 4$, the four congruence classes are

$$E_0 = \{ ..., -8, -4, 0, 4, \quad 8, 12, ... \},$$
$$E_1 = \{ ..., -7, -3, 1, 5, \quad 9, 13, ... \},$$
$$E_2 = \{ ..., -6, -2, 2, 6, 10, 14, ... \} \text{ and}$$
$$E_3 = \{ ..., -5, -1, 3, 7, 11, 15, ... \}.$$

It is easy to see that every integer is in one and only one of these.

6. Algebra of congruences

Congruence relations have a property which distinguishes them from other equivalence relations on Z, namely that two congruences mod m can be added, subtracted or multiplied like ordinary equations. By this we mean the following: *If x, x', y, y' are integers such that $x \equiv x'$ mod m and $y \equiv y'$ mod m, then (i) $x + y \equiv x' + y'$ mod m, (ii) $x - y \equiv x' - y'$ mod m and (iii) $xy \equiv x' y'$ mod m.*

Proof. We are given that $x - x'$ and $y - y'$ are both multiples of m. Therefore $(x + y) - (x' + y') = (x - x') + (y - y')$

is a multiple of m, which proves (i). The proof of (ii) is just as easy. Finally $xy - x'y' = x(y - y') + (x - x')y'$, showing that this too is a multiple of m, and this proves (iii).

For an account of many interesting applications of congruences, the reader is referred to the book *The Higher Arithmetic*, by H. Davenport (Hutchinson's University Library).

Example 41. *Tests for divisibility.* Applying (iii) repeatedly we can see that if $x \equiv y \bmod m$, then $x^n \equiv y^n \bmod m$, for any positive integer n. For example $10 \equiv 1 \bmod 9$, hence $10^n \equiv 1^n \equiv 1 \bmod 9$, for any n. Now let $x = b_r b_{r-1} \ldots b_1 b_0$ be any positive integer, written in ordinary decimal notation. Then $x = b_0 + 10b_1 + \ldots + 10^r b_r$, and by the rules we have proved, $x \equiv b_0 + 1.b_1 + \ldots + 1.b_r \bmod 9$, i.e. x has the same residue mod 9 as $b_0 + b_1 + \ldots + b_r$, the sum of the digits of x; in particular, x is divisible by 9 if and only if this sum is divisible by 9. A similar test for divisibility by 3 works because $10 \equiv 1 \bmod 3$. See also Exercise 9, p. 22.

Example 42. The *binomial theorem* for a positive exponent p gives

$$(1) \qquad (1 + t)^p = 1 + pt + \frac{p(p-1)}{1.2} t^2 + \ldots + t^p,$$

t being a variable. The coefficient of $t^r (0 \leqslant r \leqslant p)$ is

$$(2) \qquad \binom{p}{r} = \frac{p(p-1) \ldots (p-r+1)}{1.2 \ldots r}.$$

Now if p is a *prime* number and if r is one of $1, 2, \ldots, p-1$, then none of the factors $1, 2, \ldots, r$ in the denominator of (2) cancels the factor p in the numerator, hence $\binom{p}{r}$ is a multiple of p, i.e. $\binom{p}{r} \equiv 0 \bmod p$. Putting this in (1) we get

$$(3) \qquad (1 + t)^p \equiv 1 + t^p \bmod p.$$

For example, $(1 + t)^5 = 1 + 5t + 10t^2 + 10t^3 + 5t^4 + t^5 \equiv 1 + t^5 \bmod 5$. Now raise both sides of (3) to the pth power. We get $(1 + t)^{p^2}$ on the left, and on the right $(1 + t^p)^p$, which by another application of (3) (putting t^p for t in (3)) is $\equiv 1 + t^{p^2} \bmod p$. Going on like this we find

$$(4) \qquad (1 + t)^{p^a} \equiv 1 + t^{p^a} \bmod p,$$

for any prime p, and any positive integer a. We use this now to prove a fact which will be needed later (Example 121, p. 64): *If p is a prime and if k, a are any positive integers, then*

$$(5) \qquad \binom{kp^a}{p^a} \equiv k \bmod p.$$

EQUIVALENCE RELATIONS

Proof. Write $h = \binom{kp^a}{p^a}$ for short. This is the coefficient of t^{p^a} in $(1 + t)^{kp^a}$. We can write $(1 + t)^{kp^a} = ((1 + t)^{p^a})^k$, so by (4), $(1 + t)^{kp^a} \equiv (1 + t^{p^a})^k$ mod p. But if we use the binomial theorem to expand $(1 + t^{p^a})^k$, we find that the coefficient of t^{p^a} is k. Therefore $h \equiv k$ mod p, which proves (5).

EXERCISES FOR CHAPTER TWO

1. Find whether the following relations are reflexive, symmetric or transitive. Which are equivalence relations? (i) $A = R$ and '$x \sim y$' means '$x - y$ is an integer'. (ii) $A = R$ and '$x \sim y$' means '$x + y$ is an integer'. (iii) $A = P$ and '$x \sim y$' means 'x and y have an ancestor in common'. (iv) A is the set of all positive integers, and '$x \sim y$' means 'x divides y' (i.e. y/x is an integer).

2. For each integer n let $X_n = \{x \in R | n \leqslant x < n + 1\}$. Prove that $\mathscr{P} = \{X_n | n \in Z\}$ is a partition of R.

3. For any real number x, let $[x]$ ('integral part of x') denote the largest integer $\leqslant x$, e.g. $[\sqrt{5}] = 2$, $[-\frac{1}{2}] = -1$, $[3] = 3$. Define a relation $\sim$ on R by letting '$x \sim y$' mean '$[x] = [y]$'. Prove that $\sim$ is an equivalence relation and that its equivalence classes are the sets X_n of Exercise 2.

4. Prove that the relation $\sim$ of Example 37, p. 18, is an equivalence relation, and that its equivalence classes are the members of $\mathscr{Q}$.

5. Write down the different congruence classes (i) mod 2; (ii) mod 3.

6. Find the residue of 2^{512} mod 5.

7. If p is a prime prove that $xy \equiv 0$ mod p implies that $x \equiv 0$ mod p or $y \equiv 0$ mod p (or both). Prove that this fails if p is not prime.

8. Let A be the congruence class of 1 mod 3, and B the congruence class of -1 mod 4. Prove that $A \cap B$ is a congruence class mod 12.

9. Let $x = b_r b_{r-1} \ldots b_1 b_0$ be a positive integer written in decimal notation. (i) Show that x is divisible by 11 if and only if $b_0 - b_1 + b_2 - \ldots$ is divisible by 11. (ii) Show that x is divisible by 12 if and only if $b_0 - 2b_1 + 4(b_2 + b_3 + \ldots)$ is divisible by 12.

CHAPTER THREE

Mappings

---

1. Mappings

The idea of a *mapping* of one set into another is one of the most fruitful of modern mathematics. Like the two other 'abstract' notions which we have already met, namely those of *set* and *relation*, it is an elementary idea, but it gives us a new power to express accurately some fundamental mathematical situations. In particular, mappings provide a way to describe connections between the elements of different sets, or of the same set.

DEFINITION. Let A, B be any sets (which might be equal). Then a *mapping θ of A into B* is determined when there is a rule or formula which assigns, to each element x of A, an element of B, called the *image of x under θ*. This element of B is usually written in one of the two standard notations

$$\theta(x), \text{ or } x\theta.$$

The first is sometimes called the functional notation. In this book we shall adopt the second notation.

We can imagine that θ is an agent which 'transforms' or 'maps' or 'projects' each element x of A to its image element $x\theta$ in B (see Figure 8).

The words *transformation, projection, substitution, operation* are often used for mappings of various kinds. Or we may think of θ as a general kind of 'function', with arguments x in A and values $x\theta$ (but in this context it would be better to use the alternative notation $\theta(x)$) in B.

C

MAPPINGS: INJECTIVE, SURJECTIVE

We shall often use Greek letters, such as θ, ϕ, ψ, α, β, etc. for mappings.

Example 43. We can define a mapping θ from the set P of all people into the set Z of all integers, by the rule: if $x \in P$, then $x\theta$ is the age of x (i.e. $x\theta$ is his age, in years, on his last birthday).

Example 44. There are in general many different mappings of one given set into another. For example we may define a second mapping ϕ of P into Z by the rule: if $x \in P$, let $x\phi$ be the height of x, measured in millimetres to the nearest millimetre.

Example 45. Ordinary *functions* provide examples of mappings. For example, let θ be the mapping of R (the set of all real numbers) into itself defined by the rule: if $x \in R$, let $x\theta = e^x$. This is a mapping of R into R. It would be more traditional to use here the 'functional notation' $\theta(x)$ instead of $x\theta$, but of course this does not alter the mapping.

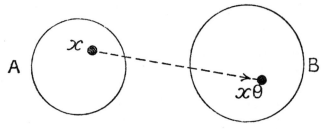

Fig. 8

2. Equality of mappings

It is convenient to use one of the notations

$$\theta : A \to B \text{ or } A \xrightarrow{\theta} B$$

to indicate that θ is a mapping of A into B. The set A is called the *domain* of θ, and B is called the *range* of θ. Notice that we do not assume that every element of the range is the image of some element of the domain (see §3).

DEFINITION. Two mappings θ and ϕ are equal (and we write $\theta = \phi$) if and only if (1) θ, ϕ have the same domain, (2) θ, ϕ have the same range, and (3) $x\theta = x\phi$ for each element x of the domain.

24

Example 46. Our definition requires that *all* of the conditions (1), (2), (3) should be satisfied. Consider the mapping $\theta:R \to R$ of Example 45. We know from elementary calculus that e^x is always *positive*, for all $x \in R$. So we could define a mapping $\phi:R \to R^+$, where R^+ is the set of all positive real numbers, by the rule $x\phi = e^x$. The only difference between θ and ϕ is that they have different ranges, i.e. (1) and (3) hold, but not (2). Therefore θ and ϕ are *not* equal*.

Mappings of finite sets. If $A = \{a_1, ..., a_m\}$ is a finite set, and B is any set, we can represent any mapping $\theta:A \to B$ by the symbol

$$\theta = \begin{pmatrix} a_1, & ..., & a_m \\ a_1\theta, & ..., & a_m\theta \end{pmatrix}$$

For example if $A = \{1, 2, 3\}$ and $B = \{a, b\}$, then $\theta = \begin{pmatrix} 123 \\ aba \end{pmatrix}$ is the mapping of A into B defined by $1\theta = a, 2\theta = b, 3\theta = a$.

If A, B are both finite, with orders m, n respectively, then there are n^m different mappings θ of A into B. For we can choose, as the image $x\theta$ of each element x of A, any one of the n elements of B; this gives $n \times n \times ... \times n = n^m$ possibilities. The $2^3 = 8$ mappings of $A = \{1, 2, 3\}$ into $B = \{a, b\}$ are $\begin{pmatrix} 123 \\ aaa \end{pmatrix}$, $\begin{pmatrix} 123 \\ aab \end{pmatrix}$, $\begin{pmatrix} 123 \\ aba \end{pmatrix}$, $\begin{pmatrix} 123 \\ abb \end{pmatrix}$, $\begin{pmatrix} 123 \\ baa \end{pmatrix}$, $\begin{pmatrix} 123 \\ bab \end{pmatrix}$, $\begin{pmatrix} 123 \\ bba \end{pmatrix}$, $\begin{pmatrix} 123 \\ bbb \end{pmatrix}$.

3. Injective, surjective, bijective; inverse mappings

The definition of a mapping does *not* require that two distinct elements x, x' of the domain should have distinct images. In Example 43, we have $x\theta = x'\theta$ whenever x, x' are two people of the same age. Also we do *not* require that every element y of the range should be the image of some element x of the domain; in the same Example, if we take $y = -3$ or $y = 1000$, then there is no x in P such that $x\theta = y$. However it is useful to have special names for those mappings which do satisfy these special requirements.

---

* This is a matter of convention. Some authors would say that θ, ϕ are equal if (1) and (3) hold, but this gives a little difficulty when defining the inverse mapping. See Example 48.

MAPPINGS

DEFINITION 1. A mapping $\theta:A \to B$ is *injective** if whenever x, x' are distinct elements of A, then $x\theta$, $x'\theta$ are distinct elements of B. Equivalently, θ is injective if $x\theta = x'\theta \Rightarrow x = x'$, for every pair of elements x, x' of A. An injective mapping is sometimes called an *injection*.

DEFINITION 2. A mapping $\theta:A \to B$ is *surjective** if for each element y of B there is at least one element x of A such that $x\theta = y$. A surjective mapping is sometimes called a *surjection*.

DEFINITION 3. A mapping $\theta:A \to B$ is *bijective* if it is both injective and surjective. A bijective mapping is sometimes called a *bijection*.

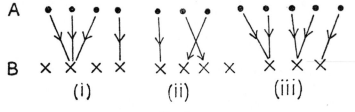

FIG. 9

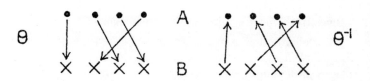

FIG. 10

Bijective mappings are particularly important. If $\theta:A \to B$ is bijective, it means that, for each $y \in B$, there is a *unique* element $x \in A$ such that $x\theta = y$ (the *uniqueness* of x comes from

---

* The terms '1−1' and 'onto' are sometimes used instead of 'injective' and 'surjective' respectively. A bijection used to be called a '1−1' correspondence'.

the fact that θ is injective, the *existence* of x from the fact that θ is surjective). Therefore we can define a mapping $\theta^{-1}:B \to A$, called the *inverse of* θ, by the rule: if $y \in B$, let $y\theta^{-1} = x$, where x is the element of A such that $x\theta = y$. Notice that θ^{-1} is defined only when θ is bijective. It is easy to see that θ^{-1} is itself bijective and that its inverse is θ again.

Example 47. Three mappings $\theta:A \to B$ are shown in Figure 9 (in each case, the row of dots is A, and the row of crosses is B). In (i) θ is neither injective nor surjective, (ii) is injective but not surjective, (iii) is surjective but not injective. Figure 10 represents a bijective mapping and its inverse. All the mappings listed at the end of the last section are surjective, except the first and last. None of them is injective.

Example 48. The mapping $\theta:R \to R$ of Example 45 is injective, because if $e^x = e^{x'}$ then $x = x'$. It is not surjective, because if y is any negative number, or zero, there is no real x such that $e^x = y$. But the mapping $\phi:R \to R^+$ of Example 46 is both injective and surjective, i.e. ϕ is bijective. Therefore there is an inverse mapping $\phi^{-1}:R^+ \to R$, and this is given by the rule: if $y \in R^+$, then $y\phi^{-1} = \log y$ (see P. J. Hilton, Differential Calculus, p. 28, in this series).

4. Product of mappings

Suppose that A, B, C are any sets, and that we have mappings $\theta:A \to B$ and $\phi:B \to C$. Then θ maps each given element x of A to its image $x\theta$, which is an element of B. If we now apply ϕ to $x\theta$ we get the element $(x\theta)\phi$ of C. So the result of applying first θ and then ϕ is a new mapping from A into C, which takes x directly to $(x\theta)\phi$. This mapping is called the *product* (or *composite*) of θ and ϕ, and is denoted $\theta\phi$. We have now made the

DEFINITION. If $\theta:A \to B$ and $\phi:B \to C$, then $\theta\phi:A \to C$ is the mapping defined by the rule: if $x \in A$, let $x(\theta\phi) = (x\theta)\phi$.

Example 49. Take $A = \{1, 2, 3\}$, $B = \{a, b\}$, $C = \{u, v, w\}$, and let $\theta = \begin{pmatrix} 123 \\ aba \end{pmatrix}$, $\phi = \begin{pmatrix} ab \\ wv \end{pmatrix}$. Then $\theta\phi = \begin{pmatrix} 123 \\ wvw \end{pmatrix}$; because, for example, θ takes 1 to a, then ϕ takes a to w, so $\theta\phi$ takes 1 to w (i.e. $w = (1\theta)\phi$). Similarly $(2\theta)\phi = v$, $(3\theta)\phi = w$.

Example 50. Take $A = B = C = \{1, 2, 3\}$, and consider the two mappings $\alpha = \begin{pmatrix} 123 \\ 231 \end{pmatrix}$, $\rho = \begin{pmatrix} 123 \\ 132 \end{pmatrix}$. Then $\alpha\rho = \begin{pmatrix} 123 \\ 321 \end{pmatrix}$, and $\rho\alpha = \begin{pmatrix} 123 \\ 213 \end{pmatrix}$, which are both mappings of A into A, but are not equal.

This 'multiplication' of mappings is very different from ordinary multiplication of numbers. In the first place the product $\theta\phi$ exists only if the range of θ is the same as the domain of ϕ. For example, there is no product $\phi\theta$ in Example 49. And Example 50 shows that even if both products $\theta\phi$ and $\phi\theta$ exist, they need not be equal. If it happens that they are equal, we say that θ and ϕ *commute*.

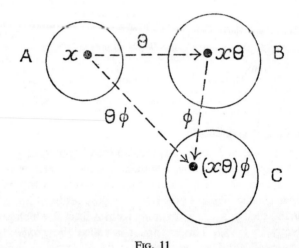

FIG. 11

Product of several mappings. Suppose that A, B, C, D are any sets, and that we have mappings $\theta:A \rightarrow B$, $\phi:B \rightarrow C$ and $\psi:C \rightarrow D$. Then we define $\theta\phi\psi:A \rightarrow D$ to be the mapping which takes a given x in A to $((x\theta)\phi)\psi$, i.e. we follow the route θ, ϕ, ψ in the diagram (Figure 12).

But we can see that this is just the same as the product of $\theta\phi$ and ψ, or equally, of θ and $\phi\psi$. This gives the *associative law for mappings*

$$(\theta\phi)\psi = \theta(\phi\psi).$$

28

Of course this law applies only if the products involved are all defined, as in the case we have shown. We can extend it to products of n mappings. If A_i $(i = 1, 2, ..., n + 1)$ are sets, and $\theta_i : A_i \to A_{i+1}$ $(i = 1, 2, ... n)$ are mappings, then $\theta_1 \theta_2 ... \theta_n : A_1 \to A_{n+1}$ is defined to be the result of applying $\theta_1, \theta_2, ... \theta_n$ in turn. This product is the same, however it is 'bracketed'. For example $(\theta_1 \theta_2)(\theta_3 \theta_4) = (\theta_1 (\theta_2 \theta_3)) \theta_4$, because both are equal to $\theta_1 \theta_2 \theta_3 \theta_4$.

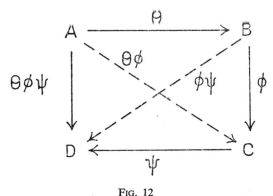

FIG. 12

Example 51. *Powers.* If θ is a mapping of A into itself, then it is possible to define the *powers* θ, θ^2, θ^3, ... of θ; they are all mappings of A into itself, θ^n being defined to be the product $\theta\theta ... \theta$ of n factors θ. For example $\theta^2 = \theta\theta$, $\theta^3 = \theta\theta\theta$. Any two powers of θ commute, because if m, n are any positive integers, then $\theta^m\theta^n$ and $\theta^n\theta^m$ are both equal to θ^{m+n}, hence $\theta^m\theta^n = \theta^n\theta^m$.

5. Identity mappings

DEFINITION. If A is any set, the *identity mapping on* A is the mapping $\iota_A : A \to A$ defined by the rule: if $x \in A$, let $x\iota_A = x$.

Clearly ι_A is bijective, and $\iota_A^{-1} = \iota_A$. With respect to mapping product, identity mappings behave rather like the number 1 in ordinary multiplication. If A, B are any sets and we have any mapping $\theta : A \to B$, then (see Example 52, below)

$$\iota_A \theta = \theta \text{ and } \theta\iota_B = \theta.$$

29

If θ is bijective, then

$$\theta\theta^{-1} = \iota_A \text{ and } \theta^{-1}\theta = \iota_B.$$

Example 52. To prove the formula $\iota_A\theta = \theta$, observe first that both $\iota_A\theta$ and θ are mappings of A into B. Then take any $x \in A$. We have $x(\iota_A\theta) = (x\iota_A)\theta = x\theta$. This proves that $\iota_A\theta = \theta$. To prove $\theta\theta^{-1} = \iota_A$, when θ is bijective, observe first that both $\theta\theta^{-1}$ and ι_A are mappings of A into A. Then take any $x \in A$. If $x\theta = y$, then by the definition of θ^{-1}, $y\theta^{-1} = x$. Therefore $x(\theta\theta^{-1}) = (x\theta)\theta^{-1} = x = x\iota_A$. Referring to the definition of equality of mappings, p. 24, we see that $\theta\theta^{-1} = \iota_A$. The formulae $\theta\iota_B = \theta$ and $\theta^{-1}\theta = \iota_B$ are proved similarly.

6. Products of bijective mappings. Permutations

Let A, B, C be any sets, and suppose that $\theta:A \to B$ and $\phi:B \to C$ are given mappings. We prove now

(i) *If θ, ϕ are both injective, then $\theta\phi$ is injective. If θ, ϕ are both surjective, then $\theta\phi$ is surjective.*

Proof. Suppose first that θ, ϕ are both injective, and that x, x' are any two distinct elements of A. Then $x\theta$, $x'\theta$ are distinct because θ is injective, hence also $(x\theta)\phi$, $(x'\theta)\phi$ are distinct because ϕ is injective. Thus we have proved that $x(\theta\phi)$, $x'(\theta\phi)$ are distinct, whenever x, x' are. This shows that $\theta\phi$ is injective.

Next suppose that θ, ϕ are both surjective, and let z be any element of C. Because ϕ is surjective there is some y in B such that $y\phi = z$, and because θ is surjective there is some x in A such that $x\theta = y$. Therefore $x(\theta\phi) = (x\theta)\phi = y\phi = z$, which proves that for every z in C there is some x in A such that $x(\theta\phi) = z$, i.e. $\theta\phi$ is surjective.

Putting together the two statements in (i) we get

(ii) *If θ, ϕ are both bijective, then $\theta\phi$ is bijective.*

There is also a partial converse to each of the statements in (i). Suppose first that $\theta:A \to B$ and $\phi:B \to C$ are such that $\theta\phi$ is injective. Then θ *must be injective.* For otherwise, there would exist distinct x, x' in A such that $x\theta = x'\theta$. Applying ϕ to both sides gives $x(\theta\phi) = x'(\theta\phi)$ — but this would contradict our assumption that $\theta\phi$ is injective. Similarly if θ, ϕ are such that $\theta\phi$ is surjective, then ϕ *must be surjective*. For take any z in C. Since $\theta\phi$ is surjective, there is some x in A such that $x(\theta\phi) = z$, i.e. $(x\theta)\phi = z$. Therefore we can always find y in B such that $y\phi = z$, namely $y = x\theta$. This proves that ϕ is surjective.

With the help of these remarks, we can prove the following lemma, which is often used to prove that a given mapping is bijective.

(iii) LEMMA. *Let* $\theta:A \to B$ *be a mapping, and suppose that there exists a mapping* $\theta':B \to A$ *which satisfies* $\theta\theta' = \iota_A$ *and* $\theta'\theta = \iota_B$. *Then* θ *is bijective and* $\theta' = \theta^{-1}$.

Proof. Since $\theta\theta' = \iota_A$, which is injective, θ is injective. Since $\theta'\theta = \iota_B$, which is surjective, θ is surjective. Hence θ is bijective. To prove $\theta' = \theta^{-1}$, calculate the product $\theta^{-1}\theta\theta'$ in two ways. First it is $(\theta^{-1}\theta)\theta' = \iota_B\theta' = \theta'$, by the results of the last section. But it is also equal to $\theta^{-1}(\theta\theta') = \theta^{-1}\iota_A = \theta^{-1}$.

Example 53. If θ, ϕ are bijections, then $(\theta\phi)^{-1} = \phi^{-1}\theta^{-1}$. This is easy to prove directly, but we shall prove it here as an application of lemma (iii). Put $\theta\phi = \alpha$, and $\phi^{-1}\theta^{-1} = \alpha'$. Then $\alpha:A \to C$ and $\alpha':C \to A$ satisfy $\alpha\alpha' = \iota_A$ and $\alpha'\alpha = \iota_C$ (for example $\alpha\alpha' = \theta\phi\phi^{-1}\theta^{-1} = \theta\iota_B\theta^{-1} = \theta\theta^{-1} = \iota_A$). Hence by lemma (iii), $\alpha = \theta\phi$ is bijective and $\alpha' = \alpha^{-1}$, i.e. $\phi^{-1}\theta^{-1} = (\theta\phi)^{-1}$.

Permutations

DEFINITION. If A is a set, then a *permutation* of A is a bijection of A into itself, i.e. a permutation of A is a mapping $\theta:A \to A$ which is both injective and surjective.

For example ι_A is a permutation of A. If θ, ϕ are permutations of A, then θ^{-1} and $\theta\phi$ are also permutations of A. These are facts which we shall need later, when we prove (Example 79, p. 43) that the set $S(A)$ of all permutations of A is a *group*.

Example 54. If $A = Z(n) = \{1, 2, ..., n\}$, then a permutation of A is a mapping $\theta = \begin{pmatrix} 1 & 2 & & n \\ a_1 & a_2 & \cdots & a_n \end{pmatrix}$, where $a_1, a_2, ..., a_n$ must be distinct elements of A, because θ is injective. Thus $a_1, a_2, ..., a_n$ must be what is called, in elementary algebra, a 'permutation' of $1, 2, ..., n$. There are n ways of choosing a_1, and then $n-1$ ways of choosing a_2 (because a_2 can be any element of A except a_1), etc., giving $n! = n(n-1) \ldots 3.2.1$ different permutations of A. We write $S(n)$ for the set of these $n!$ permutations. $S(1)$ has one element $\begin{pmatrix} 1 \\ 1 \end{pmatrix}$. $S(2)$ has two elements, $\begin{pmatrix} 12 \\ 12 \end{pmatrix}$ (the identity map on $\{1, 2\}$) and $\begin{pmatrix} 12 \\ 21 \end{pmatrix}$. The six elements of $S(3)$ are

$$\begin{pmatrix} 123 \\ 123 \end{pmatrix}, \begin{pmatrix} 123 \\ 231 \end{pmatrix}, \begin{pmatrix} 123 \\ 312 \end{pmatrix}, \begin{pmatrix} 123 \\ 132 \end{pmatrix}, \begin{pmatrix} 123 \\ 321 \end{pmatrix}, \begin{pmatrix} 123 \\ 213 \end{pmatrix}.$$

Example 55. Let a be a fixed integer, and let θ_a be the mapping of Z into itself defined by the rule: if $x \in Z$, let $x\theta_a = x + a$. Then θ_a is a permutation of Z, or as we could also put it, $\theta_a \in S(Z)$. For if x, x' are distinct integers, $x + a$ and $x' + a$ are distinct, so that θ_a is injective. And if y is any integer, there is always an integer x such that $x + a = y$, hence θ_a is surjective. Notice that $\theta_a^{-1} = \theta_{-a}$, and if $a, b \in Z$, $\theta_a\theta_b = \theta_{a+b}$.

The mapping $\theta : R \to R$ defined by $x\theta = e^x$ is not a permutation of R, because (Example 48, p. 27) it is not surjective.

7. Similar sets

Mappings play the important rôle, in all parts of mathematics, of *comparing* different sets. The simplest type of comparison is that given in the next definition.

DEFINITION. Sets A, B are said to be *similar* if there exists a bijective mapping $\theta : A \to B$.

We write $A \simeq B$ to denote that A, B are similar. This is a relation between sets, and it has the formal properties of an equivalence relation. If A is any set, then $A \simeq A$, because there is always the identity mapping $\iota_A : A \to A$, which is bijective. Then if $A \simeq B$ it follows that $B \simeq A$, because if there is a bijective mapping $\theta : A \to B$, then θ^{-1} is a bijective mapping of B into A. Finally if $A \simeq B$ and $B \simeq C$, then $A \simeq C$. For if $\theta : A \to B$ and $\phi : B \to C$ are given bijective mappings, then $\theta\phi : A \to C$ is bijective, by §6(ii).

Counting. To say that a finite set A has order n, means that we can *count* the elements $a_1, a_2, \ldots, a_n$ of A and get the 'answer' n. Now this process of counting just consists in setting up a bijective mapping

$$\theta : \{1, 2, \ldots, n\} \to A,$$

32

in which $1\theta = a_1$, $2\theta = a_2$, ... $n\theta = a_n$. Conversely if A is any set and if θ is any bijection of $\{1, 2, ..., n\}$ into A, then A must have order n. For the elements $1\theta, 2\theta, ..., n\theta$ of A are all *distinct*, since θ is injective, and they are *all* the elements of A, since θ is surjective. Write $Z(n)$ for the set $\{1, 2, ..., n\}$. What we have just said shows that *A has order n if and only if A is similar to Z(n)*. Thus similarity has a very simple meaning for finite sets, namely *finite sets A, B are similar if and only if they have the same order*.

Similarity is the basis of a theory of 'infinite numbers', or 'infinite cardinals'; we consider that two infinite sets A, B have the same 'number' of elements, if they are similar. A surprising number of the infinite sets which occur 'naturally' are, in fact, similar to one of the sets Z or R; but Z is not similar to R, so that these are infinite sets with different 'numbers' of elements*.

Example 56. In Example 48, p. 27, we found a bijective mapping $\phi:R \to R^+$, which shows that R is similar to R^+. So we can have the situation that a set is similar to a proper subset of itself. For a *finite* set A this is impossible, because any *proper* subset B of A has smaller order than A.

Example 57. Let $B = \mathscr{B}(X)$ be the set of all subsets of a set X of order n. We shall prove that B has order 2^n (see also Example 35, p. 17) by proving that B is similar to the set M of all mappings of X into the two-element set $U = \{0, 1\}$. (M has order 2^n, see §2. Any other two-element set would do as well as U.) To do this, we must define a bijective mapping $\theta:B \to M$. If $S \in B$, i.e. if $S \subseteq X$, define a mapping $\chi_S:X \to U$ by the rule: for $x \in X$, let $x\chi_S = 1$ if $x \in S$, and $x\chi_S = 0$ if $x \notin S$. Then our mapping $\theta:B \to M$ is given as follows: if $B \in B$, let $S\theta = \chi_S$. It is easy to verify that θ is bijective, and we leave this as an exercise to the reader.

---

* For a proof that Z is not similar to R, and a discussion of infinite cardinals, see S. Swierczkowski, *Sets and Numbers*, in this series.

MAPPINGS

EXERCISES FOR CHAPTER THREE

1. Write down all the mappings of the set $B = \{a, b\}$ into $A = \{1, 2, 3\}$. How many of these are injective? Find a formula for the number of injective mappings of a set of order r into a set of order s ($r \leqslant s$).

2. Which of the following mappings $\theta : Z \to Z$ are (a) injective, (b) surjective? In each case $x\theta$ is given for any $x \in Z$. (i) $x\theta = x^2$; (ii) $x\theta = -x$; (iii) $x\theta = \frac{1}{2}x$ if x is even, $x\theta = 0$ if x is odd; (iv) $x\theta = 2x + 1$.

3. Let S be the set of all real numbers $x \geqslant 0$, and define $\theta : S \to S$ by $x\theta = x^2$. Prove θ is bijective, and find its inverse. Now define $\phi : R \to R$ by $x\phi = x^2$. Is ϕ bijective?

4. Let a, b be fixed real numbers, and define $\theta : R \to R$ by the formula: $x\theta = ax + b$. Prove that θ is bijective if and only if $a \neq 0$, and give a formula for θ^{-1} in this case.

5. Write $\theta_{a,b}$ for the mapping of the preceding Exercise. Prove that $\theta_{a,b}\theta_{c,d} = \theta_{ac, bc+d}$, for any real numbers a, b, c, d. Under what circumstances do $\theta_{a,b}$ and $\theta_{c,d}$ commute?

6. If $\theta : A \to B$ and $\phi : B \to A$ satisfy $\phi\theta = \iota_B$, does it follow that θ is bijective? (Try A, B as in Exercise 1.) Prove that if $\alpha = \theta\phi$, then $\alpha^2 = \alpha$.

7. Let β be the permutation of $Z(n) = \{1, 2, ..., n\}$ which takes 1 to 2, 2 to 3, ..., $n-1$ to n, n to 1. Prove that $\beta^n = \iota_{Z(n)}$.

8. Show that if A is a finite set, then any mapping $\theta : A \to A$ which is injective is also surjective. Find an example in this Chapter which shows this to be false for an infinite set A. Construct an example of a mapping $\theta : A \to A$ which is surjective but not injective. Can this be done with A finite?

9. If A, B, C are any sets, prove (i) $A \times B \simeq B \times A$, and (ii) $(A \times B) \times C \simeq A \times (B \times C)$.

10. Show that the set of all even integers is similar to the set of all integers.

11. If X, Y are two disjoint sets, prove that $\mathscr{B}(X \cup Y) \simeq \mathscr{B}(X) \times \mathscr{B}(Y)$. (Define $\theta : \mathscr{B}(X \cup Y) \to \mathscr{B}(X) \times \mathscr{B}(Y)$ by the rule: if $S \in \mathscr{B}(X \cup Y)$, i.e. if $S \subseteq X \cup Y$, let $S\theta = (S \cap X, S \cap Y)$. Prove θ is bijective.)

12. If $A \simeq B$, prove that $S(A) \simeq S(B)$. (Let θ be a bijective mapping of A into B. Now define $\pi : S(A) \to S(B)$ as follows: if $\alpha \in S(A)$, i.e. if α is a permutation of A, let $\alpha\pi = \theta^{-1}\alpha\theta$. Check that $\alpha\pi$ *is* a permutation of B, and then verify that π is bijective.)

CHAPTER FOUR

Groups

---

1. Binary operations on a set

If x and y are ordinary numbers, there are various ways of combining or 'operating with' x and y to give another number, for example we may form their sum $x + y$, or difference $x - y$, or product xy. These are three examples of *binary operations* on the set R of all real numbers.

Modern algebra is largely concerned with the abstract properties of operations like this. In general, a binary operation ω on an arbitrary set A is nothing more than a mapping of the set $A \times A$ into A, so that ω maps each *pair* (x, y) of elements x, y of A, to some element $(x, y)\omega$ of A. For example, the operation of addition is the mapping $\alpha:R \times R \to R$ which maps (x, y) to the element $(x, y)\alpha = x + y$ of R. Similarly, subtraction is the mapping $\beta:R \times R \to R$ defined by the rule: $(x, y)\beta = x - y$. Multiplication* is the mapping $\gamma:R \times R \to R$ defined by the rule: $(x, y)\gamma = xy$.

A general binary operation $\omega:A \times A \to A$ can be thought of as a kind of generalized 'multiplication' or 'addition' on A. This point of view is emphasized if we use a notation such as $x \bigcirc y$ in place of $(x, y)\omega$; if we do this, we can in fact dispense with the symbol ω altogether, and refer simply to 'the binary operation $\bigcirc$'. With this notation our definition runs as follows.

---

* Division is not a binary operation on R, according to our definition. For the quotient x/y is not defined when $y = 0$, i.e. there is no mapping $\delta:R \times R \to R$ defined by the rule: $(x, y)\delta = x/y$, because for example $(1, 0)\delta$ would not exist.

DEFINITION. Let A be any set. Then a *binary operation* $\bigcirc$ *on* A is a mapping of $A \times A$ into A, which maps each pair (x, y) of elements of A to an element $x \bigcirc y$ of A.

Additive and multiplicative notations. It is very common, particularly in group theory, to use one of the familiar notations $x + y$ or xy, in place of $x \bigcirc y$, even when x, y are not numbers and $\bigcirc$ is not one of the ordinary arithmetical operations. We say then that we are using the *additive*, or *multiplicative*, *notation* for the operation in question. We would then usually refer to the operation itself as 'the operation $+$' (if additive notation is being used) or 'the operation . ' (if multiplicative notation is being used — we are in some embarrassment here, since the notation xy uses no symbol between x and y! In this book we shall keep to the convention that if a binary operation is called . , then it maps the pair (x, y) to xy, and *not* to $x.y$).

Example 58. If A is a small finite set, then a binary operation on A can be represented by its *multiplication table*. The table below represents a binary operation $\bigcirc$ on $A = \{a, b, c\}$, for which $a \bigcirc a = b$, $a \bigcirc b = c$, $a \bigcirc c = b$, $b \bigcirc a = b$, etc.

	a	b	c
a	b	c	b
b	b	a	c
c	a	c	c

If we decided to use the additive notation for this operation, we should write $a + a = b$, $a + b = c$, etc. In multiplicative notation, these would be written $aa = b$, $ab = c$, etc.

Example 59. Let $S(A)$ be the set of all permutations of a given set A (p. 31). Then the product of mappings defines a binary operation on $S(A)$, i.e. if $\theta, \phi \in S(A)$, we take $\theta \bigcirc \phi$ to be $\theta\phi$, which again belongs to $S(A)$. Notice that we are using the convention of multiplicative notation here.

Example 60. Let $\mathscr{B}(X)$ be the set of all subsets of a given set X (Example 25, p. 11). If $A, B \in \mathscr{B}(X)$, i.e. if A, B are subsets of X, then $A \cap B$ and $A \cup B$ also belong to $\mathscr{B}(X)$. In this way we have two binary operations $\cap$ and $\cup$ on $\mathscr{B}(X)$. The *difference* $A - B$ (p. 8) gives another binary operation on $\mathscr{B}(X)$.

2. Commutative and associative operations

In this section and the next we describe some special kinds of operation, which are relevant to the definition of a group (see §4).

Commutative operations

DEFINITION. If $\bigcirc$ is a binary operation on a set A, and if x, y are elements of A, we say that x and y *commute* (with respect to $\bigcirc$) if $x \bigcirc y = y \bigcirc x$. The operation $\bigcirc$ is called *commutative* if it satisfies the *commutative law*

$$(1) \qquad x \bigcirc y = y \bigcirc x$$

for all $x, y \in A$, i.e. if every pair of elements of A commute.

Example 61. In additive or multiplicative notations, (1) reads

$$x + y = y + x,$$
or
$$xy = yx,$$

respectively. Since these are both true when x, y are ordinary numbers, it means that the operations of ordinary addition and multiplication are both commutative. The operation of subtraction is not commutative, because if we take $x \bigcirc y = x - y$, then (1) reads $x - y = y - x$, which is not true for all x, y.

Example 62. We saw in Example 50, p. 27 an example of two mappings α, ρ, which are in fact permutations of the set $\{1, 2, 3\}$, which do not commute. This shows that the binary operation of Example 59, is not, in general, commutative.

Example 63. The operations $\cap$ and $\cup$ of Example 60 are both commutative, since $A \cap B = B \cap A$ and $A \cup B = B \cup A$, for any $A, B \in \mathscr{B}(X)$.

Associative operations

DEFINITION 2. A binary operation $\bigcirc$ on a set A is called *associative* if it satisfies the *associative law*

$$(2) \qquad (x \bigcirc y) \bigcirc z = x \bigcirc (y \bigcirc z)$$

for all $x, y, z \in A$.

Example 64. In additive or multiplicative notations, (2) reads

$$(x + y) + z = x + (y + z),$$

or
$$(xy)z = x(yz),$$

respectively. Both of these are true when x, y, z are ordinary numbers, which shows that the operations of ordinary addition and multiplication are both associative. Subtraction of numbers is not associative, because if we take $x \bigcirc y = x - y$, (2) reads $(x - y) - z = x - (y - z)$, which is not true for all x, y, z.

Example 65. The operation of mapping product on $S(A)$ (Example 59) is associative, by the associative law for mappings (p. 28). Thus it is possible to have an associative operation, which is not commutative (Example 62).

Example 66. The operations $\cap$ and $\cup$ of Example 60 are both associative (Example 27, p. 12).

Example 67. The operation of Example 58, p. 36, is neither commutative nor associative. For $a \bigcirc b \neq b \bigcirc a$, and therefore it is not commutative. It is not associative; for example $(a \bigcirc a) \bigcirc b \neq a \bigcirc (a \bigcirc b)$ (the left side is $b \bigcirc b = a$, and the right side is $a \bigcirc c = b$).

General associative law. Suppose for the moment that $\bigcirc$ is any binary operation on a set A. If we want to work out a product

(3) $$x_1 \bigcirc x_2 \bigcirc \ldots \bigcirc x_n$$

of n factors $x_1, x_2, \ldots, x_n \in A$, we must do this as a succession of products of *pairs* of elements of A, and we can show by putting in brackets, how this is to be done. For example there are five ways of working out a product of four factors x, y, z, w, viz. $x \bigcirc (y \bigcirc (z \bigcirc w))$, $x \bigcirc ((y \bigcirc z) \bigcirc w)$, $(x \bigcirc y) \bigcirc (z \bigcirc w)$, $((x \bigcirc y) \bigcirc z) \bigcirc w$, $(x \bigcirc (y \bigcirc z)) \bigcirc w$. In general, these would all be different. But if $\bigcirc$ is *associative*, then they will all be the same.

THEOREM (*General associative law*). *If $\bigcirc$ is an associative binary operation on A, and if $x_1, x_2, \ldots, x_n$ are given elements of A, then the product* (3) *has the same value, however it is 'bracketed'.*

This means that when $\bigcirc$ is associative, there is no need to put brackets into a product like (3).

Proof of the general associative law. We prove this by induction on n. For $n = 1, 2$ there is no problem, because there is only one way of working out (3) in these cases. For $n = 3$ there are two bracketings, viz. $(x_1 \bigcirc x_2) \bigcirc x_3$ and $x_1 \bigcirc (x_2 \bigcirc x_3)$, and of course these *are* equal, because we are assuming that $\bigcirc$ satisfies the associative law. Now assume $n > 3$, and as induction hypothesis, that products of fewer than n factors are independent of bracketing. In whatever way (3) is worked out, the last step will be to make a product of the form

$$P_r = (x_1 \bigcirc, \dots \bigcirc x_r) \bigcirc (x_{r+1} \bigcirc \dots \bigcirc x_n),$$

for some value of r in the range $1, 2, \dots, n-1$. In P_r it is assumed that $x_1 \bigcirc \dots \bigcirc x_r$ and $x_{r+1} \bigcirc \dots \bigcirc x_n$ have already been worked out — by the induction hypothesis, these products are independent of the way they were bracketed. Now we complete our proof by showing that $P_1 = P_2 = \dots = P_{n-1}$, for we know that (3), however it is bracketed, is equal to one of the P_r. For any r in the range $1, 2, \dots, n-2$ put $x = x_1 \bigcirc \dots \bigcirc x_r$, $y = x_{r+1}$ and $z = x_{r+2} \bigcirc \dots \bigcirc x_n$. Then (2) gives $(x \bigcirc y) \bigcirc z = x \bigcirc (y \bigcirc z)$, i.e. $P_{r+1} = P_r$. Hence $P_{n-1} = P_{n-2} = \dots = P_2 = P_1$, as required.

Powers. Let $\bigcirc$ be an associative binary operation on a set A. If we take $x_1 = x_2 = \dots = x_n$ in (3), and if we write $x = x_1$, then (3) is called the *nth power $x \bigcirc x \bigcirc \dots \bigcirc x$* of x. In multiplicative notation (and also often in the general notation) this is written x^n. In additive notation the nth power is written nx, because if we take A to be the set of all real numbers and $+$ as ordinary addition, then $x + x + \dots + x$, with n terms x, is in fact the ordinary product nx. In the case of a *general* associative operation $+$ we still use the notation nx, but of course this should no longer be thought of as a product.

Just as for powers of a mapping (p. 29) we have, for any positive integers m, n, the 'index law'.

$$(4) \qquad x^m \bigcirc x^n = x^{m+n},$$

because $x^m \bigcirc x^n$ is just a product of $m + n$ factors x. It follows that $x^m \bigcirc x^n = x^n \bigcirc x^m$, i.e. *any two powers of a given element commute*, although we are not assuming that $\bigcirc$ is a commutative operation.

Example 68. Powers, beyond the second, cannot even be defined for a general binary operation. Let $\bigcirc$ be the operation of Example 58, p. 36. If we try to define a^3, we have two choices, viz. $(a \bigcirc a) \bigcirc a$ and $a \bigcirc (a \bigcirc a)$, and these are not equal (they are $b \bigcirc a = b$ and $a \bigcirc b = c$ respectively). This indicates the kind of difficulty encountered in studying non associative operations.

Example 69. In additive notation, the index law (4) becomes

$$mx + nx = (m + n)x.$$

Example 70. If the operation $\bigcirc$ is *both* associative and commutative, then the factors in the product (3) can be permuted in any way without altering its value. In particular it is easy to prove, for any $x, y \in A$ and any positive integer n, that $(x \bigcirc y)^n = x^n \bigcirc y^n$, for each side is simply a product of n factors equal to x with n factors equal to y. But if $\bigcirc$ is not commutative, this fails (see Exercise 9).

3. Units and zeros

Let A be any set, and $\bigcirc$ a binary operation on A.

DEFINITION 1. Any element e of A which satisfies

$$(1) \qquad e \bigcirc x = x \bigcirc e = x,$$

for all x in A, is called a *unit element* (or a *neutral element*) for the operation $\bigcirc$.

There is no need for such a unit element to exist, as we shall see (Example 72, below). But if A does contain a unit element e for the operation $\bigcirc$, then it is *unique*, i.e. we have the

THEOREM. *If e, f are both unit elements for $\bigcirc$, then $e = f$.*

Proof. We put $x = f$ in (1) and get $e \bigcirc f = f$. But f satisfies $f \bigcirc x = x \bigcirc f = x$ for all x in A, so putting $x = e$, we get in particular $e \bigcirc f = e$. Thus $e = e \bigcirc f = f$.

Example 71. The number 1 is a unit element for ordinary multiplication, because $1x = x1 = x$ for all x in R. There is also a unit element for the ordinary operation of addition, but this is not 1, but the number 0. For if we take $x \bigcirc y$ to mean $x + y$, then (1) becomes $e + x = x + e = x$, which is satisfied, for all x in R, by $e = 0$.

Example 72. There is no unit for the ordinary operation of subtraction. For this would have to be a number e such that $e - x = x - e = x$ for all x in R, which is impossible. The operation $\bigcirc$ of Example 58 has no unit.

Example 73. The identity permutation ι_A is a unit element for the operation of mapping product on $S(A)$ (Example 59, p. 36), for $\iota_A \theta = \theta \iota_A = \theta$ for all θ in $S(A)$ (see p. 29).

40

Zeros. A unit element e 'leaves alone' any element x by which it is 'multiplied'. As an opposite extreme, we define next elements which 'swallow up' every other element.

DEFINITION 2. Any element n of A which satisfies

$$(2) \qquad n \bigcirc x = x \bigcirc n = n,$$

for all x in A, is called a *zero element* (or *annihilator*) for the operation $\bigcirc$.

Just as for units, there is no need for a given binary operation to have a zero element, but if it does have one, then this is unique, i.e. we have the

THEOREM. *If n, p are both zero elements for $\bigcirc$, then $n = p$.*

Proof. We put $x = p$ in (2) and get $n \bigcirc p = n$. But p satisfies $p \bigcirc x = x \bigcirc p = p$ for all x in A, so putting $x = n$, we get in particular $n \bigcirc p = p$. Thus $n = n \bigcirc p = p$.

Example 74. The number 0 is a zero element for ordinary multiplication, because $0x = x0 = 0$ for all x in R. But there is no zero element for the ordinary operation of addition, for this would have to be a number n such that $n + x = x + n = n$ for all x in R, and this is impossible.

Example 75. The subsets X, $\varnothing$ of a given set X are the unit and zero elements, respectively, for the operation $\cap$ on $\mathscr{B}(X)$ (Example 60, p. 36). For $X \cap A = A \cap X = A$, and $\varnothing \cap A = A \cap \varnothing = \varnothing$, for all subsets A of X. For the operation $\cup$, these rôles are reversed; X is the zero and $\varnothing$ the unit element.

4. Gruppoids, semigroups and groups

In this section we shall define the most elementary *algebraic structures*. An algebraic structure consists of a *set* (usually assumed to be non-empty), together with one or more *operations* on this set. For example, if G is any set and $\bigcirc$ is any binary operation on G, then the pair $(G, \bigcirc)$ is called a *gruppoid*. This is the first type of algebraic structure we shall consider.

DEFINITION 1. A gruppoid is a pair $(G, \bigcirc)$, where G is a non-empty set, and $\bigcirc$ is a binary operation on G.

If G is a finite set, we say that $(G \bigcirc)$ is a *finite gruppoid*, with *order* $|G|$. Otherwise $(G, \bigcirc)$ is an *infinite gruppoid*.

Example 76. A gruppoid $(G, \bigcirc)$ consists of the two 'components', a *set G*, and a *binary operation* $\bigcirc$ on G. There can be many different gruppoids with the same set. For example if R is the set of all real numbers, then $(R, +)$, $(R, -)$ and $(R, .)$ (here . stands for ordinary multiplication) are three different gruppoids.

If $\cap$ in the operation of Example 58, p. 36, then $(A, \bigcirc)$ is a finite gruppoid of order 3. All the examples of binary operations which we have given, provide examples of gruppoids.

DEFINITION 2. A *semigroup*[*] is a gruppoid $(G, \bigcirc)$ whose operation $\bigcirc$ is associative, i.e. which satisfies the associative law

G1 $(x \bigcirc y) \bigcirc z = x \bigcirc (y \bigcirc z)$ for all x, y, z in G.

Example 77. $(R, +)$ and $(R, .)$ are semigroups, but not $(R, -)$ (see Example 64, p. 38). $(\mathscr{B}(X), \cap)$ and $(\mathscr{B}(X), \cup)$ are both semigroups (Example 66, p. 38).

DEFINITION 3. A *group* is a semigroup $(G, \bigcirc)$ which satisfies, in addition to **G1**, two further conditions

G2 G has a *unit element e*, i.e.

$$e \bigcirc x = x \bigcirc e = x \text{ for all } x \text{ in } G, \text{ and}$$

G3 Every element of G has an *inverse*, i.e. for each x in G there is an element $\hat{x}$ in G, called the inverse of x, such that

$$x \bigcirc \hat{x} = \hat{x} \bigcirc x = e.$$

Thus *a group is a set G, together with a binary operation $\bigcirc$ on G, such that* **G1**, **G2** *and* **G3** *hold*. The conditions **G1, G2, G3** are called the *group axioms;* the theorems of *group theory* are those which can be deduced from these axioms.

Every group $(G, \bigcirc)$ is of course also a semigroup, and every semigroup is a gruppoid. So gruppoids are the most 'general' of the three types of structure we have defined. Any theorems about arbitrary gruppoids would have very wide applications, because examples of gruppoids occur in mathematics at every turn. Unfortunately very few non-trivial theorems about gruppoids have been found, so that, although very general,

---

[*] Sometimes called a *monoid*.

42

gruppoid theory is also very dull. The introduction of the associative law **G1** makes more progress possible, but it is the combination of all three assumptions **G1**, **G2**, **G3** which gives group theory a unique position; it is general enough to draw ideas and interest from all parts of mathematics, but at the same time special and detailed enough to inform these ideas with a characteristic expression which comes from its own 'personality' For the rest of this book we shall discuss some of these ideas or concepts of group theory.

Abelian groups. The commutative law is not one of the axioms of group theory. But if it happens that a group $(G, \bigcirc)$ does satisfy this law, i.e. if

G4 $x \bigcirc y = y \bigcirc x$ for all x, y in G,

then $(G, \bigcirc)$ is called an *Abelian* (or *commutative*) group. (This adjective Abelian is taken from the name of the Norwegian mathematician N. H. Abel (1802–1829).) Thus an Abelian group is a gruppoid which satisfies **G1**, **G2**, **G3** and **G4**.

5. Examples of groups

Example 78. *One-element groups.* If a group $(G, \bigcirc)$ has only one element, then, since it must have a unit e, $G = \{e\}$, and the operation $\bigcirc$ is completely defined by $e \bigcirc e = e$. Conversely if we take any one-element set $G = \{e\}$ and define $\bigcirc$ in this way, then it is easy to check **G1**, **G2**, **G3** and verify that $(G, \bigcirc)$ is a group.

Example 79. *Symmetric groups.* Let A be any given set, and $S(A)$ the set of all permutations of A. If we denote the binary operation giving the mapping product $\theta\phi$ by . , then $(S(A), .)$ is a group, called the *symmetric group* on A. We verify this by checking the group axioms.

G1 holds, since mapping product is associative (p. 28).

G2 holds, taking $e = \iota_A$ (Example 73, p. 40).

G3 holds, because if $\theta \in S(A)$, then $\theta^{-1} \in S(A)$ and satisfies $\theta\theta^{-1} = \theta^{-1}\theta = \iota_A$ (p. 30), i.e. we can take $\hat{\theta} = \theta^{-1}$.

When A is the set $\{1, 2, ..., n\}$, we write $S(A) = S(n)$. The group $(S(n), .)$ is finite, of order $n!$ (Example 54, p. 31); it is called the *symmetric group of degree n*. The 'multiplication table' for $(S(3), .)$ is given on p. 78. As we have already seen (Example 62, p. 37), this is not an Abelian group.

Example 80. *The additive group of integers.* Let Z be the set of all integers, and let $+$ denote ordinary addition. Then $(Z, +)$ is a group. **G1** holds, i.e. $(x + y) + z = x + (y + z)$ for all integers x, y, z. **G2** holds if we take $e = 0$. **G3** holds if we take $\hat{x} = -x$. This is an Abelian group, because it satisfies also the commutative law $x + y = y + x$ for all $x, y \in Z$.

Example 81 *The additive group of reals.* Let R be the set of all real numbers and $+$ the ordinary operation of addition. Then $(R, +)$ is an Abelian group, just as in the last example. The unit element is again 0, and the 'inverse' $\hat{x}$ of x is $-x$.

Example 82. *The multiplicative group of reals.* $(R, .)$ is a semigroup, but not a group. For **G1** holds, $(xy)z = x(yz)$ for all $x, y, z \in R$, and in fact **G2** holds, taking $e = 1$. But there is no number 0 satisfying $0\hat{0} = 0\hat{0} = 1$, so that 0 does not have an inverse, and **G3** fails. However, if we take instead of R the set R^* of all *non-zero* real numbers, then $(R^*, .)$ is a group (notice that if $x \in R^*, y \in R^*$ then $xy \in R^*$, i.e. multiplication does define a binary operation on R^*). Because now **G1**, **G2** hold just as before, and also **G3**, since for any $x \in R^*$ we take $\hat{x} = 1/x$. $(R^*, .)$ is an Abelian group.

Example 83. A group $(G, \bigcirc)$ which has a zero element n must be a one-element group. For if x is any element of $G, x = x \bigcirc e = x \bigcirc (n \bigcirc \hat{n}) = (x \bigcirc n) \bigcirc \hat{n} = n \bigcirc \hat{n} = e$. Thus $G = \{e\}$.

Additive and multiplicative notations for groups. It is usual in group theory to use either the additive or multiplicative notations for the group operation $\bigcirc$. There are some standard conventions, which we now describe.

1. *Additive notation.* In this notation we write

$x + y$ for $x \bigcirc y$ (and call this the *sum* of x and y),

0   for the unit element e (but notice, this is *not* a zero element),

$- x$ for the inverse $\hat{x}$ of an element x of G,

$x - y$ for $x \bigcirc \hat{y}$,

nx for the nth power $x \bigcirc x \bigcirc ... \bigcirc x, n$ a positive integer.

The notation nx is extended to arbitrary integers by writing $0x = 0$ and $(-n)x = n(-x) = -(nx)$ (see theorem (v), next section).

2. *Multiplicative notation.* In this notation we write

  xy for $x \bigcirc y$ (and call this the *product* of x and y),
  e or 1   for the unit element,
  x^{-1} for the inverse $\hat{x}$,
  x^n for the nth power $x \bigcirc x \bigcirc ... \bigcirc x$, n a positive integer.

The notation x^m is extended to arbitrary integers by writing $x^0 = e$ and $x^{-n} = (x^{-1})^n = (x^n)^{-1}$ (see theorem (v), next section).

The additive notation is often, but not exclusively, reserved for Abelian groups (e.g. $(Z, +)$, Example 80).

6. Elementary theorems on groups

We collect in this section the simplest theorems about groups. Like all the theorems of group theory, they are proved using only the axioms **G1**, **G2** and **G3**. It does not matter which convention of notation we use, and we shall use multiplicative notation. Everything could be 'translated' into additive notation if we preferred (see Example 84, below).

Let $(G, .)$ be a group, e its unit element. By the theorem on p. 40, *e is unique*. Let us write x^{-1} for the inverse $\hat{x}$ of a given element x of G; axiom **G3** says that such an element exists and satisfies (a) $x^{-1}x = e$, and (b) $xx^{-1} = e$. The next theorem shows that *the inverse of x is unique*, in fact that x^{-1} is the only element of G which satisfies either of the conditions (a), (b).

(i) THEOREM. *If $x \in G$ and if a is any element of G such that $ax = e$, then $a = x^{-1}$. Similarly if b is any element such that $xb = e$, then $b = x^{-1}$.*

We shall prove this as a consequence of the more general

(ii) THEOREM. *If $x, y \in G$ then there is one and only one element a of G such that $ax = y$, namely $a = yx^{-1}$. Similarly there is one and only one element b of G such that $xb = y$, namely $b = x^{-1}y$.*

Proof. If $ax = y$, multiply on the right by x^{-1}, giving $(ax)x^{-1} = yx^{-1}$. But $(ax)x^{-1} = a(xx^{-1})$ (by **G1**) $= ae$ (by **G3**) $= a$ (by **G2**); therefore $ax = y$ implies $a = yx^{-1}$. Conversely if $a = yx^{-1}$, then $ax = (yx^{-1})x = y(x^{-1}x) = ye = y$. We leave to the reader the corresponding task of proving that $xb = y$ if and only if $b = x^{-1}y$. To prove theorem (i) we now have only to put $y = e$ in theorem (ii).

(iii) THEOREM. *If $x \in G$ then $(x^{-1})^{-1} = x$.*

Proof. Since $xx^{-1} = e$ by **G3**, $x = (x^{-1})^{-1}$ (replace a in (i) by x, and x by x^{-1}).

(iv). THEOREM. *If $x, y \in G$ then $(xy)^{-1} = y^{-1}x^{-1}$.*

Proof. Replace a in (i) by $y^{-1}x^{-1}$ and x by xy. Since $(y^{-1}x^{-1})(xy) = (y^{-1}(x^{-1}x))y$ by the general associative law (p. 38), this product is equal to $(y^{-1}e)y = y^{-1}y = e$. Hence (i) gives $y^{-1}x^{-1} = (xy)^{-1}$.

This result extends easily to a product of n factors, $(x_1x_2 \ldots x_n)^{-1} = x_n^{-1} \ldots x_2^{-1}x_1^{-1}$. Taking $x_1, x_2, \ldots, x_n$ all equal to x we get a proof of the following

(v). THEOREM. *If $x \in G$ and n is any positive integer, $(x^n)^{-1} = (x^{-1})^n$.* This justifies the notation, which we have already introduced, by which we write x^{-n} for $(x^n)^{-1}$; it is the same as $(x^{-1})^n$.

Example 84. In additive notation, theorem (ii) would read: 'If $x, y \in G$ there is one and only one element a of G such that $a + x = y$, namely $a = y + (-x)$ (or $y - x$). Similarly, there is one and only one element b of G such that $x + b = y$, namely $b = (-x) + y$'.

Example 85. The inverse of e is e. (Take $a = x = e$ in theorem (i); $ee = e$ by axiom **G2**.)

Example 86. Each element x commutes with its inverse, because axiom **G3** says $xx^{-1} = x^{-1}x = e$.

Example 87. *General index law.* If m, n are positive integers

(vi) $$x^m x^n = x^{m+n} \text{ for any } x \in G,$$

by the associative law (p. 39); this still holds if either m, n is zero, by our convention $x^0 = e$. We show now that (vi) holds for arbitrary integers m, n. As we have seen, it does hold if (a) $m \geqslant 0$ *and* $n \geqslant 0$. Next we prove it in the case (b) $m \geqslant 0$ *and* $m + n \geqslant 0$. We have only to consider the case $n < 0$, say $n = -s\ (s > 0)$. Then by (a) $x^{m+n}x^s = x^{m+n+s} = x^m$. Now multiply both sides on the right by $(x^s)^{-1} = x^n$ and we get (vi). A similar proof holds in case (c) $n \geq 0$ and $m + n \geq 0$. Suppose finally (d) *No two of m, n, $m + n$ are ≥ 0.* Then certainly at least two of $-n$, $-m$, $-n-m$ are positive, and we can apply one of the previous cases to prove $x^{-n}x^{-m} = x^{-n-m}$. Now take inverses of both sides, and we get (vi). So (vi) holds for any integers m, n. By a similar discussion we may prove also (vii) $(x^m)^n = x^{mn}$ *for any $x \in G$, and any integers m, n.*

EXERCISES FOR CHAPTER FOUR

1.   Show that there are n^{n^2} different binary operations on a set A of order n. Find all 16 binary operations on $A = \{a, b\}$. How many are (i) commutative, (ii) associative, (iii) have unit elements or (iv) have zero elements?

2.   Define $\bigcirc$ on R by the rule: if $x, y \in R$ then $x \bigcirc y = xy + 1$. Show that this is commutative but not associative. Does it have unit or zero elements?

3.   If $\bigcirc$ is a binary operation on a set A, and if an element z is both a unit and a zero for $\bigcirc$, show that $A = \{z\}$.

4.   An element e which satisfies $e \bigcirc x = x$ for all x in A is a 'left unit' for $\bigcirc$. Show that left units are not necessarily unique and (hence) they may not be unit elements, as follows. Let $A = \{a, b\}$ and define $\bigcirc$ by $a \bigcirc a = b \bigcirc a = a, a \bigcirc b = b \bigcirc b = b$.

5.   Define right units similarly, and show that if $\bigcirc$ has a right unit e and a left unit f, then $e = f$ and this is a unit element.

6.   Make similar definitions and prove results analogous to the above for left and right zeros of a binary operation.

7.   Let $M(A)$ denote the set of all mappings of a set A into itself. Show that $(M(A), .)$ is a semigroup, where . denotes mapping product. Is this a group?

8.   Let $(G, .)$ by a semigroup and x, y elements of G which commute. Show that x^m, y^n commute, where m, n are any positive integers.

9.   Taking α, ρ as in Example 50, p. 27, show that $(\alpha\rho)^2 \neq \alpha^2\rho^2$.

10.   Define mappings α, β of Z into itself as follows: for any $x \in Z$, $x\alpha = 2x$; if x is even, $x\beta = \frac{1}{2}x$, while if x is odd, $x\beta = 0$. Show that $\alpha\beta = \iota$, the identity mapping on Z, but $\beta\alpha \neq \iota$. Prove that if $(G, .)$ is any semigroup with unit element e, and if $a, b \in G$ are such that $ab = e$, then $(ba)^2 = ba$. Verify this in the example just given.

11.   Define a binary operation $\bigcirc$ on R by $x \bigcirc y = x + y + xy$. Prove that $(R, \bigcirc)$ is a semigroup, with unit element 0. If R' is the set of real numbers $x \neq -1$, show that $(R', \bigcirc)$ is a group.

12.   Show that $(\mathscr{B}(X), \oplus)$ is an Abelian group, where $\mathscr{B}(X)$ is the set of all subsets of a set X, and $\oplus$ is defined in Exercise 7, p. 13.

13. Let $(G, .)$ be a semigroup with unit element e, and say that $x \in G$ is *invertible* if there exists some $y \in G$ such that $xy = yx = e$. Prove that if x, x' are invertible, so is xx', and show that $(G^*, .)$ is a group, where G^* is the set of all invertible elements. Find the order of this group when $(G, .) = (Z, .)$.

14. If x, y are elements of a group and $x^2 = y^2 = (xy)^2 = e$, prove that x and y commute.

15. Let G be the set of all mappings $\theta_{a,b}$ of Exercises 4, 5, p. 34, such that $a \neq 0$. Prove that $(G, .)$ is a non-Abelian group, where . denotes mapping product.

CHAPTER FIVE

Subgroups

---

1. Subsets closed to an operation

Let G be a set, and $\bigcirc$ a binary operation on G.

DEFINITION. A non-empty subset H of G is said to be *closed to* $\bigcirc$ if, for every pair (x, y) of elements of H, also $x \bigcirc y$ belongs to H; i.e. if H satisfies the condition

S1 $\qquad x, y \in H \Rightarrow x \bigcirc y \in H.$

Example 88. Let Z be the set of all integers, and Y the subset consisting of all integers $x \geq 0$. Then Y is closed to both the operations of addition and multiplication, because if $x, y \in Y$ then both $x + y$ and xy belong to Y. But Y is not closed to subtraction (for example $0, 1 \in Y$ but $0 - 1 \notin Y$).

If H is closed to $\bigcirc$, we can use $\bigcirc$ as a binary operation on H,* so that $(H, \bigcirc)$ is a gruppoid which is a part of the gruppoid $(G, \bigcirc)$; we say sometimes that H is a *sub-gruppoid* of $(G, \bigcirc)$. If $\bigcirc$ is associative as operation on G, then it is associative as operation on H (for if $(x \bigcirc y) \bigcirc z = x \bigcirc (y \bigcirc z)$ for all x, y, z in G, then certainly this holds for all x, y, z in H!); similarly if $\bigcirc$ is commutative on G. In particular if H is a sub-gruppoid of a *semigroup* $(G, \bigcirc)$, then $(H, \bigcirc)$ is itself a semigroup. But the last Example shows that if $(G, \bigcirc)$ is a *group* and H is closed to $\bigcirc$, then $(H, \bigcirc)$ need not be a group — for $(Z, +)$ is a group, Y is closed to $+$, but $(Y, +)$ is not a group since it does not contain the inverse $-y$ of each element $y \in Y$.

---

*It would be more accurate, but perhaps tedious, to use separate symbols for $\bigcirc$ as operation on G, which is a mapping $G \times G \to G$, and for $\bigcirc$ as operation on H, which is a mapping $H \times H \to H$.

2. Subgroups

DEFINITION. Let $(G, \bigcirc)$ be a group, and H a non-empty subset of G. Then H is called a *subgroup of* $(G, \bigcirc)$ if

S1 $x, y \in H \Rightarrow x \bigcirc y \in H$, and

S2 $x \in H \Rightarrow \hat{x} \in H$.

Thus H has to be closed to $\bigcirc$, and also to the 'operation' of taking inverses.

THEOREM. *If H is a subgroup of $(G, \bigcirc)$, then $(H, \bigcirc)$ is a group.*

Proof. $(H, \bigcirc)$ satisfies the associative law **G1**, as we have seen. Then it also satisfies **G2**, because in fact we have the

LEMMA. *Any subgroup H of $(G, \bigcirc)$ contains the unit element e of G.*

To prove this lemma, take any $x \in H$. By **S2**, $\hat{x} \in H$. Then using **S1**, $x \bigcirc \hat{x} = e \in H$.

Finally $(H \bigcirc)$ satisfies **G3**, by condition **S2**. This proves the theorem.

The two conditions **S1**, **S2** can be put as a single condition, as follows: *a non-empty subset H of G is a subgroup if and only if*

S $x, y \in H \Rightarrow x \bigcirc \hat{y} \in H$.

First suppose H is a subgroup, i.e. satisfies **S1** and **S2**. Let $x, y \in H$. By **S2**, $\hat{y} \in H$; now by **S1**, $x \bigcirc \hat{y} \in H$, this verifies **S**. Conversely suppose H satisfies **S**. Take any $x \in H$, and put $y = x$ in **S**, which gives $x \bigcirc \hat{x} = e \in H$. Put e, x for x, y in **S**, and we find $e \bigcirc \hat{x} = \hat{x} \in H$; this shows H satisfies **S2**. To verify that H satisfies **S1**, take any $x, y \in H$. We know $\hat{y} \in H$, and we know also (Theorem (iii), p. 46) $(\hat{\hat{y}}) = y$. So put $x, \hat{y}$ for x, y in **S**, and get $x \bigcirc y \in H$.

In additive and multiplicative notations, **S** becomes
$$x, y \in H \Rightarrow x - y \in H, \text{ and}$$
$$x, y \in H \Rightarrow xy^{-1} \in H,$$
respectively.

Example 89. If H is a subgroup of $(G, \bigcirc)$, and $x \in H$, then all the powers x^n (n any integer) belong to H.

If $(G, \bigcirc)$ is Abelian, so is $(H, \bigcirc)$.

Example 90. G itself is a subgroup of any group $(G, \bigcirc)$. So is the one-element set $\{e\}$ (check **S1** and **S2** — the only values x and y can have are $x = e$ and $y = e$!).

Example 91. Z is a subgroup of $(R, +)$ (check **S**: if $x, y \in Z$, then also $x - y \in Z$). $(Z, +)$ is the group of Example 80, p. 44. The set R^+ of all positive real numbers is a subgroup of $(R^*, .)$. (Example 82, p. 44) (check **S**: if x, y are positive, so is xy^{-1}). In Example 88, Y is not a subgroup of $(Z, +)$, because Y does not satisfy **S2**, although it does satisfy **S1**.

Example 92. Let m be any positive integer, and define $mZ = \{\ldots, -2m, -m, 0, m, 2m, \ldots\}$, the set of all multiples of m. Then mZ is a subgroup of $(Z, +)$ (if $x, y \in mZ$, clearly $x - y \in mZ$).

Example 93. The following are all the subgroups of the symmetric group $S(3)$ (see p. 78): $S(3)$, $\{\iota, \alpha, \beta\}$, $\{\iota, \rho\}$, $\{\iota, \sigma\}$, $\{\iota, \tau\}$, $\{\iota\}$. Notice that they all contain the unit element ι.

Example 94. Let $\mathscr{S}$ be any set o subgroups of $(G, .)$. Then $J = \bigcap \mathscr{S}$ (see p. 11) is a subgroup. For let $x, y \in J$. This means, $x, y \in H$ for each member H of $\mathscr{S}$. Since each H is a subgroup, $xy^{-1} \in H$. But then $xy^{-1} \in H$ for each $H \in \mathscr{S}$, so $xy^{-1} \in J$, i.e. J satisfies **S**. In particular if H, K are subgroups of $(G, .)$, then so is $H \cap K$. In general $H \cup K$ is not a subgroup — take $H = \{\iota, \rho\}$, $K = \{\iota, \sigma\}$ in the last Example.

3. Subgroup generated by a subset

From now on we shall use *multiplicative* notation for a general group unless we explicitly indicate the contrary; also we shall follow a standard practice and speak of 'the group G' instead of 'the group $(G, .)$'.

Let X be a non-empty subset of a group G, not necessarily a subgroup of G. By a *word in X*, we shall mean any element of G which can be expressed in the form

$$(1) \qquad u_1^{m_1} u_2^{m_2} \ldots u_f^{m_f},$$

where f is any positive integer, $u_1, u_2, \ldots, u_f$ are any elements of X (not necessarily distinct), and $m_1, m_2, \ldots, m_f$ are any integers.

The *set of all words in X* will be denoted gp X.

Example 95. gp X is the set of all elements of G which you can get, starting with the elements of X and then forming products and inverses any number of times. For example gp $\{x, y\}$ includes $x, y, xy, \; yx, \; xyx, x^{-1}y, \; yx^{-1}, \; xy^{-1}, \; xy^{-1}x, \; x^{-1}y^{-1}, \; xyx^{-1}$ etc.

51

Example 96. There are infinitely many expressions of type (1), but of course they do not all give different elements of G. Taking $X = \{\alpha, \rho\}$ in the symmetric group $S(3)$ we soon find that every element of $S(3)$ is a word in X, and in many ways. For example $\beta = \alpha^2 = \alpha^{-1}\rho^2 = \alpha \rho \, \alpha^{-1}\rho^{-1} = \rho \, \alpha \, \rho = \rho^{17}\alpha^{-2}\rho^9\alpha^9$ etc.

THEOREM. *For any non-empty subset X of G, gp X is a subgroup of G.*

Proof. Let $x = u_1^{m_1} \dots u_f^{m_f}$ and $y = v_1^{n_1} \dots v_g^{n_g}$ be any elements of gp X ($u_1, \dots, u_f, v_1, \dots, v_g \in X$). Then

$$xy^{-1} = u_1^{m_1} \dots u_f^{m_f} v_g^{-n_g} \dots v_1^{-n_1},$$

which is also a word in X. Hence gp X satisfies the condition **S** of p. 50.

DEFINITION. gp X is called the subgroup of G *generated by X*. If gp $X = G$, i.e. if every element of G is a word in X, we say G is *generated by X*, and that X is a *set of generators* for G.

Example 97. $S(3)$ is generated by $\{\alpha, \rho\}$. For every element of $S(3)$ can be written as a word in $\{\alpha, \rho\}$, e.g. $\iota = \alpha^0$, $\sigma = \alpha\rho$, etc. Equally, this group is generated by $\{\sigma, \tau\}$, as the reader may confirm. Any group G is generated by G itself. If H is a subgroup of G, then $H = $ gp H.

Example 98. If G has a finite set X of generators, then G is said to be *finitely generated*. For example Z (i.e. $(Z, +)$) is generated by $\{1\}$, since every element of Z has the form $m1$ for some integer m, and $m1$ is a word, additively written, in the set $\{1\}$. See Example 99, below.

4. Cyclic groups

DEFINITION. A group G is called *cyclic* if there is an element x of G such that $G = $ gp $\{x\}$. Any such element x is called a *generator of G*.

Each element of gp $\{x\}$ has the form (1) of §3, where now all of $u_1, \dots, u_f$ must be equal to x, i.e. it is a power x^m of x. So the cyclic group gp $\{x\}$ consists of the elements

$$(1) \qquad \dots, x^{-2}, x^{-1}, x^0 = e, x, x^2, \dots,$$

which need not all be distinct. Since $x^m x^n = x^{m+n} = x^n x^m$ for any integers m, n (Example 87, p. 46), *every cyclic group is Abelian*.

Example 99. In additive notation the elements (1) are

$$\ldots, -2x, -x, 0x = 0, x, 2x, \ldots.$$

For example the subgroup mZ of Z (Example 92) is cyclic and is generated by m. It is also generated by $-m$.

Example 100. If e is the unit element of a group G, then gp$\{e\}$ is the one-element subgroup $\{e\}$. For $e^n = e$ for any integer n.

There are two kinds of cyclic group gp $\{x\}$, corresponding to the two possibilities that the elements (1) are (a) all distinct, or (b) not all distinct.

(a) *Infinite cyclic groups.* If all the elements (1) are distinct, i.e. if $x^a \neq x^b$ whenever $a \neq b$, gp $\{x\}$ is infinite, and we say also that x is an *element of infinite order*.

(b) *Finite cyclic groups.* If there exist distinct a, b such that $x^a = x^b$, we can assume $a > b$ and deduce $x^{a-b} = x^{b-b} = x^o = e$. This shows that there exist positive integers p such that $x^p = e$; now define m to be the *smallest positive integer such that $x^m = e$, m is called the *order of x*. Now we prove the

LEMMA. *If m is the order of x, and if a, b are any integers, then $x^a = x^b$ if and only if $a \equiv b \bmod m$.*

Proof. First we observe that $x^{qm} = (x^m)^q = e^q = e$, for any integer q. So if $a \equiv b \bmod m$, there exists an integer q such that $a = b + qm$, hence $x^a = x^{b+qm} = x^b\, x^{qm} = x^b e = x^b$. Conversely, suppose that $x^a = x^b$. As before we find $x^p = e$, where $p = a - b$. Let $p = qm + s$, where s is the residue of p mod m, so that s is one of $0, 1, \ldots, m-1$. Now $e = x^p = x^{qm+s} = x^{qm}x^s = x^s$. If $s > 0$, this contradicts the fact that m is the smallest positive integer such that $x^m = e$. Therefore $s = 0$, $p = a - b = qm$, which shows that $a \equiv b \bmod m$. This proves the lemma. Since each integer n is congruent mod m to exactly one of the integers $0, 1, \ldots, m-1$, gp $\{x\}$ is the set of m elements.

(2) $x^o = e, x, x^2, \ldots, x^{m-1}.$

Notice that the order of $x = m = |\text{gp } \{x\}|$.

Example 101. From the lemma, putting $b = 0$, we have that if x has order m and a is any integer, $x^a = e$ *if and only if a is a multiple of m*. Also notice $x^{-1} = x^{m-1}$. The element α of $S(3)$ (p. 78) has order 3, and gp $\{\alpha\} = \{\iota, \alpha, \alpha^2\} = \{\iota, \alpha, \beta\}$. The unit element of any group has order 1. Every non-zero element x of Z has infinite order, for $mx = 0$ only if $m = 0$. Table 2 (p. 79) shows a cyclic group of order 6.

In a given group G, its subgroups are 'features' which we look for. One way to get a subgroup is to take any *subset* X of G and find the subgroup gp X which it generates — for example each element x of G gives us a cyclic subgroup gp $\{x\}$. But there is another very natural source of subgroups, based on the idea of a group 'acting' on a set, which we define in the next section.

5. Groups acting on sets

Let A be a set, and $(G, .)$ a group written with multiplicative notation. It often happens that we can define a 'product' in which each pair (a, x) of elements, $a \in A$ and $x \in G$, combine to give an element of A, which we shall write $a * x$. (This is very similar to a binary operation; $*$ could be defined as a mapping of $A \times G$ into A.)

DEFINITION. If A is any set and $(G, .)$ any group, we say *G acts on A by the product* $*$ if, for each pair (a, x) of elements $a \in A$ and $x \in G$ there is defined an element $a * x$ of A, in such a way that for all $a \in A$ and for all $x, y \in G$ the following axioms hold:

A1 $\qquad a * e = a$, and

A2 $\qquad (a * x) * y = a * (xy)$.

We can think of each element x of G 'acting' on the set A, in the sense that it changes $a \in A$ to another element $a * x \in A$. Then **A1** says that the unit element e 'acts' by leaving each element of A unaltered, and **A2** says that if first x, and then y act, the combined effect is the same as the action of xy.

Example 102. Take any set A, and let $G = S(A)$ (Example 79, p. 43). For any $a \in A$, $\theta \in S(A)$ define $a * \theta$ to be $a\theta$, the image of a under θ. Then $S(A)$ acts on A with this product; the axioms **A1** ($a * \iota_A = a$) and **A2** (($a * \theta) * \phi = a * (\theta\phi)$, for θ, $\phi \in S(A)$) are just the definitions of ι_A and of $\theta\phi$, respectively.

Example 103. *Transforms.* Let G be any group, and take $A = G$, i.e. we shall make G act on itself, as follows: if a, $x \in G$ define $a * x = x^{-1}ax$. This element $x^{-1}ax$ is usually written a^x, and called the *transform*† *of a by x.* **A1** holds, $a^e = e^{-1}ae = a$ for all $a \in G$. Also **A2** holds, because $(a^x)^y = y^{-1}(x^{-1}ax)y = y^{-1}x^{-1}axy = (xy)^{-1}a(xy) = a^{xy}$, i.e. $(a * x) * y = a * (xy)$.

Example 104. We can make it act on the set $B = \mathscr{B}(G)$ of all subsets of G, as follows. If $U \in B$ i.e. if $U \subseteq G$, and if $x \in G$, define $U * x$ to be the set $Ux = \{ux | u \in U\}$. (For example take $U = \{\iota, \alpha\}$, subset of the group $S(3)$. Then $U\sigma = \{\iota\sigma, \alpha\sigma\} = \{\sigma, \tau\}$, which is again a subset of $S(3)$.) The axioms now read **A1**: $Ue = U$, and **A2**:$(Ux)y = U(xy)$, for any subset U of G, and any elements x, $y \in G$; these are easy to verify.

Example 105. Now we shall make G act on the same set B as in the last Example, but in a different way, by defining $U * x$ to be the set $U^x = \{u^x | u \in U\}$. (In the example $U = \{\iota, \alpha\}$, $x = \sigma$ we should now have $U^\sigma = \{\iota^\sigma, \alpha^\sigma\} = \{\sigma^{-1}\iota\sigma, \sigma^{-1}\alpha\sigma\} = \{\iota, \beta\}$.) The axioms now read **A1**: $U^e = U$, and **A2**:$(U^x)^y = U^{xy}$; these follow readily from what was proved in Example 103.

Example 106. Let H be a subgroup of G. Then H acts on G, if we define, for $a \in G$ and $x \in H$, $a * x = ax$ (i.e. the ordinary product of a and x as elements of G). **A1** and **A2** follow at once from the group axioms — in particular **A2** from the associative law in G.

6. Stabilizers

DEFINITION. Let G be a group which acts on a set A by a product $*$, and let a be a given element of A. Then the set

$$G_a = \{x \in G | a * x = a\}$$

is called the *stabilizer* of a.

G_a is the set of all elements of G which leave a unchanged, or 'stable'. We prove now a fundamental, but elementary

THEOREM. G_a *is a subgroup of* G.

Proof. We shall verify conditions **S1** and **S2** (p. 50) for G_a. **S1**. If $x, y \in G_a$, it means $a * x = a$ and $a * y = a$. Hence $a * xy = (a * x) * y = a * y = a$, i.e. $xy \in G_a$, as required.

---

† Do not confuse this with the *power* a^n ($n \in Z$).

S2. If $x \in G_a$, so that $a * x = a$, then also $a * x^{-1} = a$, hence $x^{-1} \in G_a$. This follows by taking $a = b$ in the

LEMMA. *If a, b are any elements of A, and x any element of G, then $a * x = b \Leftrightarrow a = b * x^{-1}$.*

Proof. From $a * x = b$ follows $(a * x) * x^{-1} = b * x^{-1}$. But $(a * x) * x^{-1} = a * (xx^{-1})$ (by **A2**) $= a * e = a$ (by **A1**). This proves that $a * x = b \Rightarrow a = b * x^{-1}$; the converse is proved similarly.

We shall return to the theory of stabilizers in the next chapter. For the moment we consider this simply as an important way of finding subgroups of groups.

Example 107. Take $G = S(n)$, the symmetric group on $A = \{1, 2, ..., n\}$, which acts on A as in Example 102. The stabilizer G_1 is the set of all permutations θ of A which map 1 to 1 — for example if $n = 3$, G_1 is the subgroup of two elements $\iota = \begin{pmatrix} 123 \\ 123 \end{pmatrix}$, $\rho = \begin{pmatrix} 123 \\ 132 \end{pmatrix}$.

Example 108. In Example 103, G acts on G by $a * x = x^{-1}ax$. The condition $a * x = a$, i.e. $x^{-1}ax = a$, is exactly the condition that a, x commute. For $x^{-1}ax = a \Leftrightarrow x(x^{-1}ax) = xa \Leftrightarrow ax = xa$. Thus the stabilizer of a, usually written $C(a)$ or $C_G(a)$ and called the *centralizer of a in G*, is the set of all elements x of G which commute with a. For example the centralizer of α in $S(3)$ is easily found to be the subgroup $\{\iota, \alpha, \beta\}$.

Example 109. The stabilizer of a subset U in Example 104 is the set of all x in G such that $Ux = U$, i.e. Ux and U must be the same *sets* — the order of elements does not matter. For example if $U = \{\iota, \tau\} \subseteq S(3)$, then τ belongs to the stabilizer of U, because $U\tau = \{\iota\tau, \tau\tau\} = \{\tau, \iota\}$, and this is the same set as U.

Example 110. The stabilizer of a subset U of G in the sense of Example 105 is called the *normalizer* (denoted $N(U)$ or $N_G(U)$) *of U in G*. It is the set of all x in G such that $U^x = U$. For example the normalizer in $S(3)$ of $U = \{\alpha, \beta\}$ is the whole group $S(3)$, because it happens that $U^\theta = U$ for all $\theta \in S(3)$, e.g. $U^\sigma = \{\alpha^\sigma, \beta^\sigma\} = \{\beta, \alpha\} = U$.

Example 111. The stabilizer of an element a of G, which is acted on by the subgroup H as in Example 106, is $\{e\}$. For the only element $x \in H$ such that $ax = a$ is $x = e$.

SUBGROUPS

EXERCISES FOR CHAPTER FIVE

1. Let X be a given set, and Y a fixed subset of X. Show that the set $\mathscr{S}$ of all subsets A of X such that $A \supseteq Y$ is closed to the operation $\cup$ on $\mathscr{B}(X)$. Is $\mathscr{S}$ closed to $\cap$?

2. Define an operation $\bigcirc$ on $A = \{1, 2, ..., n\}$, in such a way that no proper subset of A is closed to $\bigcirc$.

3. Let $(G, .)$ be any Abelian group, and let H be a non-empty subset of G which is closed to ., i.e. H satisfies **S1** but not necessarily **S2**. Show that the set $H^* = \{xy^{-1} | x \in H, y \in H\}$ is a subgroup of G.

4. Show that the set H of all non-zero integers is not a subgroup of the group $(R^*, .)$, but that it does satisfy **S1**. Find the subgroup H^* (see Exercise 3) in this case.

5. Let m, n be positive integers. Prove that $mZ \cap nZ = lZ$, where l is the least common multiple of m, n.

6. H is a subgroup of a group $(G, .)$, and x, y are elements of G. Prove that if *any two* of x, y, xy are in H, then so is the third.

7. If H, K are subgroups of a group G and if $H \cup K$ is also a subgroup, prove that either $H \subseteq K$ or $K \subseteq H$.

8. Prove that $S(3)$ is generated by $\{\alpha, \rho\}$, and also by $\{\sigma, \tau\}$.

9. If X is a subset of G such that $xy = yx$ for every pair (x, y) of elements of X, prove that gp X is Abelian.

10. Let m, n be any positive integers. Regarding these as elements of the group Z, prove that gp $\{m, n\} = $ gp $\{d\}$, where d is the highest common factor of m and n.

11. Show that gp X is the smallest subgroup of G which contains X, i.e. if H is any subgroup of G and $H \supseteq X$, prove that $H \supseteq$ gp X.

12. Show that gp $X = \bigcap \mathscr{S}_X$, where $\mathscr{S}_X$ is the set of all subgroups of G which contain X.

13. Find the orders of all the elements of $A(4)$ (p. 80). Also show that the permutation β of Exercise 7, p. 34, has order n.

14. If x is a generator of a cyclic group G, show that also x^{-1} generates G. Find all the generators of $(Z, +)$. Find all the generators of a cyclic group $G = $ gp $\{x\}$ of order 8.

15. If G is a *finite* group, prove that any non-empty subset H of G which satisfies **S1** (p. 50) also satisfies **S2**, i.e. is a subgroup. (Prove that every element of G has finite order.)

16. Find the centralizers of all the elements of $S(3)$.

17. Find the normalizer in $S(4)$ of the set $\{\beta, \beta^{-1}\}$, where $\beta = \begin{pmatrix} 1234 \\ 2341 \end{pmatrix}$.

18. If G acts on a set A and if $a \in A$, $x \in G$, prove $G_{a*x} = x^{-1}G_a x$.

19. When G acts on $\mathscr{B}(G)$ as in Example 104, p. 55, show that the stabilizer of a subgroup H of G, regarding H as element of $\mathscr{B}(G)$, is H itself.

CHAPTER SIX

Cosets

---

1. The quotient sets of a subgroup

Throughout this section G is a group, written with multiplicative notation, and H is a subgroup of G. We are going to show that H determines two *partitions* of G, called the right and left *quotient sets*, respectively, of G by H. This fact is a very striking feature of group theory; there is no counterpart to it in the theory of general gruppoids, or even semigroups.

We need first some notation (already introduced in Example 104, p. 55). If U is any subset of G and x any element of G let $Ux = \{ux | u \in U\}$. This is again a subset of G; if U is a finite set $\{u_1, ..., u_n\}$ then $Ux = \{u_1 x, ..., u_n x\}$. Similarly define $xU = \{xu | u \in U\}$. In general, $Ux \neq xU$. But from the associative law in G we find at once that $(Ux)y = U(xy)$, $(xU)y = x(Uy)$ and $x(yU) = (xy)U$ for any $x, y \in G$, so we need not use brackets in these products. Also $eU = Ue = U$.

DEFINITION. Let H be a subgroup of a group G. Any set Hx, with $x \in G$, is called a *right coset* of H in G. The set of all right cosets Hx, $x \in G$, is called the *right quotient set* of G by H, denoted G/H. Similarly any set xH is called a *left coset* of H in G, and the set of all left cosets xH, $x \in G$, is called the *left quotient set* of G by H, denoted $G \backslash H$.

Now let x, y be elements of G. Write $u = yx^{-1}$, which is equivalent to $y = ux$. Since y belongs to Hx if and only if $y = ux$ for some element u of H, we have

(i) LEMMA. $y \in Hx \Leftrightarrow yx^{-1} \in H$.

Next we define a relation $\sim$ on G as follows: if $x, y \in G$, '$x \sim y$' shall mean '$yx^{-1} \in H$'.

(ii) LEMMA. $\sim$ is an equivalence relation on G.

Proof. **E1** holds, for if $x \in G$, then $xx^{-1} = e \in H$, because H is a subgroup. **E2** holds, for if $x \sim y$, i.e. $yx^{-1} \in H$, then also $xy^{-1} = (yx^{-1})^{-1} \in H$ by **S2**, so that $y \sim x$. Finally **E3** holds, for if $x \sim y$ and $y \sim z$, i.e. yx^{-1} and zy^{-1} both belong to H, then **S1** shows that $zx^{-1} = (zy^{-1})(yx^{-1}) \in H$, so that $x \sim z$.

Now consider the *equivalence class E_x* of a given element x of G. By definition (p. 17) this is the set of all $y \in G$ such that $x \sim y$, i.e. such that $yx^{-1} \in H$. Thus lemma (i) shows that E_x *is exactly the right coset Hx*, and we can apply the fundamental theorem on equivalence relations (p. 17), and obtain the

THEOREM. *The set $G/H = \{Hx \,|\, x \in G\}$ is a partition of G. If x, y are elements of G then $Hx = Hy \Leftrightarrow yx^{-1} \in H$.*

In exactly the same way, but using the equivalence relation defined by '$x^{-1}y \in H$', we have the analogous result for the left quotient set: $G \backslash H = \{xH \,|\, x \in G\}$ *is a partition of G. If x, y are elements of G then $xH = yH \Leftrightarrow x^{-1}y \in H$.*

Example 112. H itself is both a right and a left coset of H, because $H = He = eH$. Putting $x = e$ in the theorem above, we have that $Hy = H$ if and only if $y \in H$; similarly $yH = H$ if and only if $y \in H$. But in general the right and left cosets Hx and xH are different, and so G/H and $G \backslash H$ are different partitions of G. We give below the right and left cosets of the subgroup $H = \{\iota, \rho\}$ of the symmetric group $S(3)$ (see p. 78). There are three

Right cosets: $H = H\iota \quad = H\rho \quad = \{\iota, \rho\}$,
$\qquad\qquad\qquad H\alpha \quad = H\tau \quad = \{\alpha, \tau\}$,
$\qquad$ and $H\beta \quad = H\sigma \quad = \{\beta, \sigma\}$.

The right quotient set is $G/H = \{\{\iota, \rho\}, \{\alpha, \tau\}, \{\beta, \sigma\}\}$. There are also three

Left cosets: $\quad H = \iota H \quad = \rho H \quad = \{\iota, \rho\}$,
$\qquad\qquad\qquad \alpha H \quad = \sigma H \quad = \{\alpha, \sigma\}$,
$\qquad$ and $\beta H \quad = \tau H \quad = \{\beta, \tau\}$.

So the left quotient set is $G \backslash H = \{\{\iota, \rho\}, \{\alpha, \sigma\}, \{\beta, \tau\}\}$.

Example 113. In additive notation a coset Hx is written $H + x$, and xH is written $x + H$. If $(G, +)$ is an Abelian group, then of course the right and left cosets coincide.

Let m be a fixed positive integer, and mZ the subgroup of $(Z, +)$ consisting of all multiples of m (Example 92, p. 51). Taking $G = Z$ and $H = mZ$, the equivalence relation $\sim$ above is given by the rule: if $x, y \in Z$, '$x \sim y$' means '$y - x \in mZ$', in other words, '$x \sim y$' is the same as '$x \equiv y \bmod m$'. So the cosets of mZ in Z are exactly the residue classes (congruence classes) of y mod m; the class E_x on p. 19 is the coset $x + mZ$ in this case the quotient set (we need not say 'right' or 'left' because $(Z, +)$ is Abelian) Z/mZ has m elements $r + mZ$ ($r = 0, 1, ..., m-1$).

2. Mappings of quotient sets

It is at first rather difficult to realise that the notation Hx for a right coset is ambiguous, because the same coset may also be written Hy, for some y different from x. In fact the theorem of the last section tells us exactly when this happens:

(i) $\qquad Hx = Hy$ *if and only if* $yx^{-1} \in H$.

Similarly for left cosets

(ii) $\qquad xH = yH$ *if and only if* $x^{-1}y \in H$.

This is particularly important when we try to define a mapping θ of G/H (or of $G\backslash H$) into some set B. We usually do this by a rule of the form

(1) $\qquad (Hx)\theta = \bar{x}$,

where $\bar{x}$ is some element of B, $\bar{x}$ being defined for each x of G. But if $Hx = Hy$, then (1) also gives $(Hy)\theta = \bar{y}$, so that unless $\bar{x} = \bar{y}$ the rule (1) is inconsistent — it does not give a well-defined image of the coset $Hx = Hy$ under θ. So whenever we define a mapping $\theta : G/H \to B$ by a rule of this type, we must verify that the rule is consistent: (1) *is consistent if and only if* $\bar{x} = \bar{y}$ *for any* x, y *of* G *such that* $yx^{-1} \in H$. A similar condition applies for mappings of $G\backslash H$.

Example 114. Take $G = S(3)$ and $H = \{\iota, \rho\}$ as in Example 112. We might try to define a mapping $\theta : G/H \to G$ by the rule $(Hx)\theta = x^2$, for any $x \in G$. But this is not consistent, for $\alpha\tau^{-1} \in H$, i.e. $H\alpha = H\tau$, but $\alpha^2 \neq \tau^2$. Thus the image under θ of the coset $\{\alpha, \tau\}$ is not well-defined.

3. Index. Transversals

THEOREM. *Let H be any subgroup of a group G. Then*

$$G/H \simeq G\backslash H.$$

In particular if either one of the two quotient sets is finite, then they both are, and they have the same order.

DEFINITION. If G/H is finite then $|G/H|$ is called the *index of H in G.*

Proof of the theorem. We have to show that G/H and $G\backslash H$ are similar sets (p. 32), i.e. that there is a bijective mapping $\theta: G/H \to G\backslash H$. We attempt to define θ by the rule

(1) $(Hx)\theta = x^{-1}H$, for any $x \in G$,

but must first verify consistency (taking $\bar{x} = x^{-1}H$). If $x, y \in G$ are such that $yx^{-1} \in H$, then also $xy^{-1} = (yx^{-1})^{-1} \in H$, i.e. $(x^{-1})^{-1}(y^{-1}) \in H$. This implies $x^{-1}H = y^{-1}H$ by 2(ii) above, so the rule (1) *is* consistent, and does define a mapping θ. To show θ is bijective, define similarly $\theta': G\backslash H \to G/H$ by the rule $(xH)\theta' = Hx^{-1}$ (and show this is consistent). It is very easy to see that $\theta\theta' = \iota_{G/H}$ and $\theta'\theta = \iota_{G\backslash H}$, and it follows that θ is bijective (Chapter 3, 6(iii), p. 31).

Example 115. If G is finite, every subgroup H has finite index, e.g. $|S(3)/H| = 3$ in Example 112. Even if G is infinite it is possible for a subgroup of G to have finite index, for example mZ has index m in Z (Example 113).

Transversals. A subset X of G is called a *right transversal* for H in G* if each right coset of H in G contains exactly one member of X. If H has finite index n in G, then $X = \{x_1, ..., x_n\}$ must have n elements, one from each right coset, and then $G/H = \{Hx_1, ..., Hx_n\}$. Similarly we may define left transversals.

Example 116. There are in general many transversals of H in G. For example the sets $\{\iota, \alpha, \beta\}$, $\{\rho, \alpha, \sigma\}$, $\{\rho, \sigma, \tau\}$ are all right transversals of H in $S(3)$ (Example 112). The set $X = \{0, 1, ..., m-1\}$ is a transversal of mZ in Z (Example 113).

---

* Or a *set of right coset representatives.*

4. Lagrange's theorem

We may notice in Example 112 (p. 59) that all the cosets of the subgroup H have the same order, i.e. they all have the same order as H (which is itself a right and a left coset). This is an example of the following general fact.

LEMMA. *Let H be a subgroup of G and x any element of G. Then $H \simeq Hx$ and $H \simeq xH$. In particular, if H is finite then*

$$|H| = |Hx| = |xH| \text{ for all } x \in G.$$

Proof. Define mappings $\theta: H \to Hx$ and $\theta': Hx \to H$ as follows: if $u \in H$ let $u\theta = ux$, and if $v \in Hx$ let $v\theta' = vx^{-1}$ (notice that if $v \in Hx$ then $vx^{-1} \in H$, by §1, lemma (i)). Verify that $\theta\theta' = \iota_H$ and $\theta'\theta = \iota_{Hx}$; this shows θ is bijective (Chapter 3, 6(iii), p. 31). Thus $H \simeq Hx$; similarly $H \simeq xH$.

Suppose now that H is a subgroup of a *finite* group G. Let $m = |H|$ and $n = |G/H|$. Then G/H is a partition of G into n cosets, each of which has m elements. Therefore the order of G is mn. This gives the following theorem, which is one of the oldest in group theory.

LAGRANGE'S THEOREM. *If H is a subgroup of a finite group G, then the order of H divides the order of G, and $|G| = |H| \, |G/H|$.*

We can write this formula $|G/H| = |G|/|H|$, which explains the notation and the term 'quotient set'.

Example 117. The order (p. 53) of an element x of a finite group G divides the order of G. For the order of x is the order of the cyclic subgroup $H = \text{gp}\{x\}$ of G.

Example 118. *If a group G has prime order p, then G is cyclic.* For let $x \in G$, $x \neq e$. The order of the subgroup $H = \text{gp}\{x\}$ must divide p, which is prime. So $|H|$ can only be 1 or p. But H has at least two elements e and x, so in fact $|H| = p$, i.e. H is the whole of G. Therefore $G = \text{gp}\{x\}$, and so is cyclic.

5. Orbits

Suppose that the group G acts on a set A, as in Chapter 5, §§5,6. This action of G defines a relation on A as follows:

if $a, b \in A$, let '$a \sim b$' mean 'there is some $x \in G$ such that $a * x = b$'.

(i) LEMMA. $\sim$ *is an equivalence relation on* A.

Proof. **E1.** If $a \in A$, then $a * e = a$, hence $a \sim a$. **E2.** If $a \sim b$, so that $a * x = b$ for some $x \in G$, then $b * x^{-1} = a$ by the lemma on p. 56, hence $b \sim a$. **E3.** If $a \sim b$ and $b \sim c$, there exist elements $x, y \in G$ such that $a * x = b$ and $b * y = c$. Then $a * (xy) = (a * x) * y = b * y = c$, and this shows that $a \sim c$.

DEFINITION. Let a be a fixed element of A. Then the equivalence class

$$O_a = \{a * x | x \in G\}$$

of a under $\sim$ is called the *orbit of a under* G.

By the fundamental theorem on equivalence relations (p. 17) the set of orbits of the elements of A is a partition of A; if A is finite and $A_1, ..., A_n$ are the different orbits, then

$$A = A_1 \cup ... \cup A_n$$

and $A_i \cap A_j = \varnothing$ whenever $i \neq j$. If A is itself a single orbit, we say A is *transitive*; this means that if a, b are any elements of A, there is always at least one element x of G such that $a * x = b$.

(ii) LEMMA. *Let a be a fixed element of A and* G_a *its stabilizer. Then for any elements* x, y *of* G, $yx^{-1} \in G_a \Leftrightarrow a * x = a * y$.

Proof. Taking $b = a * y$ in the lemma of Chapter 5, §6 (p. 56), we have $a * x = a * y \Leftrightarrow a = (a * y) * x^{-1} = a * (yx^{-1}) \Leftrightarrow yx^{-1} \in G_a$. This lemma allows us to prove a theorem which connects the orbit O_a of a given element a of A, with its stabilizer G_a.

THEOREM. *For any fixed element a of A,* $G/G_a \simeq O_a$. *In particular if G is finite,* $|O_a| = |G/G_a|$, *hence by Lagrange's theorem*

$$|O_a| = |G|/|G_a|.$$

Proof. We may define a mapping $\theta : G/G_a \to O_a$ by the rule $(G_a x)\theta = a * x$, for all $x \in G$. For the lemma above tells us that this rule is consistent in the sense of §2. We want to prove that θ is bijective. First θ is clearly surjective, by the definition of O_a. And if $x, y \in G$ are such that $(G_a x)\theta = (G_a y)\theta$, then we have $a * x = a * y$ which implies $yx^{-1} \in G_a$, i.e. $G_a x = G_a y$, again by lemma (ii). So θ is injective, and this proves the theorem.

Example 119. *Conjugacy.* G acts on G (Examples 103, 108) by $a * x = x^{-1}ax$. Elements $a, b \in G$ are called *conjugate in G* if $a \sim b$ in this sense, i.e. if there is some $x \in G$ such that $x^{-1}ax = b$, and the orbits O_a are called the *conjugacy classes* of G. For example $S(3)$ has three conjugacy classes, $A_1 = \{\iota\}$, $A_2 = \{\alpha, \beta\}$ and $A_3 = \{\rho, \sigma, \tau\}$. The stabilizer of a is its centralizer $C(a)$, hence if G is finite the order of the conjugacy class of a is $|G/C(a)|$; notice that this divides the order of G.

Example 120. Let r, n be fixed positive integers, $r \leqslant n$, and let B_r be the set of all subsets of order r of the set $A = \{1, 2, ..., n\}$. The symmetric group $S(n)$ acts on B_r, if we define $U * \theta$ to be the set $U\theta = \{u\theta | u \in U\}$, for any $U \in B_r$ and $\theta \in S(n)$. Now B_r is *transitive*, because if $U = \{u_1, ..., u_r\}$ and $V = \{v_1, ..., v_r\}$ are any subsets of A of order r, it is possible to find a permutation θ of A such that $u_1\theta = v_1, ..., u_r\theta = v_r$, and so $U\theta = V$. So B_r is the orbit of any one of its members, e.g. of $U = \{1, 2, ..., r\}$. The stabilizer G_U of U is the set of all permutations

$$\theta = \begin{pmatrix} 1 & 2 & ... & r & r+1 & ... & n \\ a_1 & a_2 & ... & a_r & a_{r+1} & ... & a_n \end{pmatrix}$$

such that $\{a_1, ..., a_r\} = U$; so $a_1, ..., a_r$ can be any of the $r!$ permutations of $1, ..., r$, and $a_{r+1}, ..., a_n$ can be any of the $(n-r)!$ permutations of $r + 1, ..., n$. Thus $|G_U| = r! (n-r)!$ and we have $|B_r| = $ |orbit of U| $= |S(n)|/|G_U| = n!/(r!(n-r)!)$ — this is the well-known fact that the number of subsets of r elements of a set of n elements, is equal to the binomial coefficient $\binom{n}{r} = n!/(r!(n-r)!)$.

Example 121. *Sylow's theorem.* There is no converse to Lagrange's theorem, i.e. if G is a group of finite order g, and if m is an integer dividing g, there may be no subgroup H of G of order m. However if p is a prime, and if p^a is the highest power of p which divides g, then G has at least one *subgroup of order p^a*. This theorem was proved by L. Sylow in 1873. Any subgroup H of G of order p^a is called a *Sylow p-subgroup of G*.

Let A be the set of all *subsets* of G of order p^a; we make G act on A as in Example 104, p. 55, i.e. if $U \in A$ and $x \in G$ we define $U * x$ to be the set Ux. Let $A_1, ..., A_n$ be the different orbits, so that

(a) $$|A| = |A_1| + ... + |A_n|.$$

Each A_i is the orbit under G of some set U_i of order p^a. If H_i is the stabilizer of U_i we have by the theorem above

(b) $|A_i| = |G|/|H_i|$ $(i = 1, ..., n)$.

If $u \in U_i$ and $x \in H_i$ then $ux \in U_i$, because $U_ix = U_i$. Keeping u fixed and letting x run over H_i we see that $uH_i \subseteq U_i$, for any $u \in U_i$. Therefore U_i is the union of all the left cosets uH_i, as u runs over U_i — of course these cosets may not all be distinct. But if the number of distinct ones is r_i, then $|U_i| = r_i|H_i|$, because distinct left cosets of H_i are disjoint, and they all have order $|H_i|$. This shows that $|H_i|$ divides $|U_i| = p^a$, and since the only factors of p^a are smaller powers of p,

(c) $|H_i| = p^{a_i}$, for some $a_i \le a$ $(i = 1, ..., n)$.

We can write $g = |G| = kp^a$, where p does not divide k. From (b) and (c) follow

(d) $|A_i| = kp^{d_i}$, where $d_i = a - a_i \ge 0$ $(i = 1, ..., n)$.

By the last Example, $|A| = \binom{kp^a}{p^a}$. We saw in Example 42, p. 21, that this $\equiv k \bmod p$. But p does not divide k, so p does not divide $|A|$, and then by (a) and (d), p does not divide

$$k(p^{d_1} + ... + p^{d_n}).$$

If $d_i > 0$ for all $i = 1, ..., n$, then p would divide this integer. So there must be some i such that $d_i = 0$. Then for this i, $a_i = a$, i.e. $|H_i| = p^a$. This proves Sylow's theorem, because H_i is a subgroup of G of order p^a.

As an example, the order of the group A(4) (p. 80) is $12 = 2^2.3$. Sylow's theorem tells us that A(4) must have at least one subgroup of order $2^2 = 4$, and at least one of order 3. In fact, A(4) has exactly one Sylow 2-subgroup V, and four Sylow 3-subgroups $\{e, a, p\}$, $\{e, b, s\}$, $\{e, c, q\}$, $\{e, d, r\}$. Notice that A(4) has no subgroup of order 6, although 6 divides $12 = |A(4)|$.

6. Normal subgroups

DEFINITION. Let H be a subgroup of a group G. Then we say that H is *normal in G*, and write $H \lhd G$, if

N $Hx = xH$ for all $x \in G$.

Thus a subgroup H is normal in G if the two quotient sets G/H, $G\backslash H$ coincide. Every subgroup H of an Abelian group G is normal in G. If we multiply **N** on the left by x^{-1} we find an equivalent condition: *H is normal in G if and only if*

N' $x^{-1}Hx = H$ for all $x \in G$.

COSETS

A third, and probably the most useful, form of this condition is:

A subgroup H of G is normal in G if and only if

N'' $u \in H, x \in G \to x^{-1}ux \in H.$

In the terminology of Example 119, H is normal in G if H contains, with any of its elements u, also all the conjugate elements to u in G. To prove that N', N'' are equivalent, observe that $x^{-1}Hx = \{x^{-1}ux \mid u \in H\}$, so N'' is itself equivalent to

N''' $x^{-1}Hx \subseteq H$ for all $x \in G$.

Clearly N' implies N'''. But if N''' holds, put x^{-1} in place of x and we have $xHx^{-1} \subseteq H$; multiplying on the left by x^{-1} and on the right by x gives $H \subseteq x^{-1}Hx$. With $x^{-1}Hx \subseteq H$, this gives N'.

Example 122. $H = \{\iota, \rho\}$ is not normal in $S(3)$; for example $\rho \in H$ but $\alpha^{-1}\rho\alpha = \sigma \notin H$. The subgroup $\{\iota, \alpha, \beta\}$ is normal in $S(3)$.

Example 123. In any group G, the subgroups G and $\{e\}$ are always normal. A group G, which is not a one-element group, and which has no normal subgroups except G and $\{e\}$, is called *simple. Any group G of prime order p is simple*, because by Lagrange's theorem *any* subgroup H of G must have order 1 or p, i.e. $H = \{e\}$ or G.

Example 124. Conjugate subgroups. In the notation of Example 103 we write $a^x = x^{-1}ax$, for any elements a, x of a group G. We prove now two identities (i) $a^x b^x = (ab)^x$, and (ii) $(a^x)^{-1} = (a^{-1})^x$, for all $a, b, x \in G$. For (i), $a^x b^x = x^{-1}axx^{-1}bx = x^{-1}abx = (ab)^x$. For (ii), $(a^x)^{-1} = (x^{-1}ax)^{-1} = x^{-1}a^{-1}x = (a^{-1})^x$. With these we can show that if H is any subgroup of G (not necessarily normal), and $x \in G$, then $H^x = x^{-1}Hx$ is a subgroup of G. For take any elements $a^x, b^x \in H^x$ $(a, b \in H)$, then $ab, a^{-1} \in H$ because H is a subgroup, hence by (i) and (ii) $a^x b^x$ and $(a^x)^{-1} \in H^x$; this proves H^x is a subgroup. H^x is called a *conjugate subgroup* to H, in G. If H is *normal* in G, it means that H coincides with all its conjugate subgroups in G.

7. Quotient groups

Let G be a group, and H any normal subgroup of G. We shall now define a binary operation $\bigcirc$ on the quotient set $G/H = G\backslash H$ by the following rule: for any cosets Hx, Hy $(x, y \in G)$ let

(1) $Hx \bigcirc Hy = Hxy.$

Our first problem is to show that this rule is consistent, i.e. if $x, x', y, y' \in G$ are such that $Hx = Hx'$ and $Hy = Hy'$, we must prove that $Hxy = Hx'y'$. This is easy, but uses the fact that H is *normal* in G, i.e. $Hz = zH$ for any $z \in G$. We have

$Hxy = Hx'y$ (since $Hx = Hx'$) $= x'Hy$ (since $Hx' = x'H$) $= x'Hy'$ (since $Hy = Hy'$) $= Hx'y'$ (since $Hx' = x'H$). If H is not normal, then rule (1) does not work (see Example 125, below).

Example 125. Take $H = \{\iota, \rho\}$, which is not normal in $S(3)$. If we try to define $\bigcirc$ on $S(3)/H$ by (1), we get contradictions. For example, if $A = H\alpha = H\tau$ and $B = H\beta = H\sigma$, then $A \bigcirc B = H\alpha \bigcirc H\beta = H\alpha\beta = H\iota = H$. But (1) also gives, $A \bigcirc B = H\tau \bigcirc H\sigma = H\tau\sigma = H\alpha = A$. So no such operation $\bigcirc$ exists.

THEOREM. *Let H be a normal subgroup of a group G, and let $\bigcirc$ be the operation defined by* (1). *Then $(G/H, \bigcirc)$ is a group.*

DEFINITION. $(G/H, \bigcirc)$ is called the *quotient group* of G by H.*

Notation. It is common to use the same notation (additive or multiplicative) for G/H as for G. So if G is written multiplicatively, we write $HxHy$ instead of $Hx \bigcirc Hy$, and if G is written additively, we write $(H + x) + (H + y)$ instead of $(H + x) \bigcirc (H + y)$. Thus the operation in the quotient group G/H is defined by

$$HxHy = Hxy, \text{ or } (H + x) + (H + y) = H + (x + y),$$

respectively.

Proof of the theorem. Using multiplicative notation, we verify **G1.** $(HxHy)Hz = (Hxy)Hz = H(xy)z = Hx(yz) = HxHyz = Hx(HyHz)$ for all $x, y, z \in G$. **G2.** The coset $H = He$ is a unit element for G/H, for $HeHx = Hx = HxHe$ for all $x \in G$. **G3.** Hx^{-1} is the inverse of Hx, for any $x \in G$. For $HxHx^{-1} = Hxx^{-1} = He = H$, similarly $Hx^{-1}\,Hx = H$.

Example 126. The subgroup mZ of $(Z, +)$ (Examples 92, 113) is normal, because $(Z, +)$ is Abelian and so all its subgroups are normal. The rule for adding two cosets is $(mZ + x) + (mZ + y) = mZ + (x + y)$. Write $\bar{x} = mZ + x$, for short. The following table gives the addition for the group $Z/3Z$, which has three elements $\bar{0}, \bar{1}, \bar{2}$. For example $\bar{2} + \bar{2} = \bar{4} = \bar{1}$, $\bar{2} + \bar{1} = \bar{3} = \bar{0}$, etc.

---

* Or *factor group.*

	$\bar{0}$	$\bar{1}$	$\bar{2}$
$\bar{0}$	0	$\bar{1}$	$\bar{2}$
$\bar{1}$	$\bar{1}$	$\bar{2}$	$\bar{0}$
$\bar{2}$	$\bar{2}$	$\bar{0}$	1

Example 127. The table on p. 80 is of a subgroup $A(4)$ of the symmetric group $S(4)$ — this is described in Example 142, p. 75. The set $V = \{e, t, u, v\}$ is easily shown to be a normal subgroup of $A(4)$. Its index, by Lagrange's theorem, is $|A(4)|/|V| = 3$, and the three cosets of V in $A(4)$ are $V, A = Va, P = Vp$. The table below gives the multiplication for the quotient group $A(4)/V$. For example $AP = VaVp = Vap = Ve = V$, etc.

	V	A	P
V	V	A	P
A	A	P	V
P	P	V	A

Notice that $A(4)/V$ is Abelian, although $A(4)$ is not.

EXERCISES FOR CHAPTER SIX

1. Show that $H = \{e, a, p\}$ is a subgroup of the group $A(4)$ of p. 80, and find all its right and left cosets in $A(4)$.

2. What are the cosets of the subgroup $\{e\}$ in a group G? What are the cosets of the subgroup G of G?

3. If H, K are subgroups of G and $x \in G$, prove $xH \cap xK = x(H \cap K)$.

4. If X is a right transversal of H in G, show that $Y = \{x^{-1} | x \in X\}$ is a left transversal.

5. If H is a subgroup of index n in G, show that a subset $\{x_1, ..., x_n\}$ of G is a right transversal of H in G if and only if $x_i x_j^{-1} \notin H$, for all $i, j \in \{1, 2, ..., n\}$ such that $i \neq j$.

6. If H is a subgroup of K, and K is a subgroup of G, and if G/K and K/H are both finite, prove that $|G/H| = |G/K| \, |K/H|$.

7. Find all the conjugacy classes (Example 119, p. 64) of $A(4)$.

8. If a subgroup H of G acts on G by the rule $a * x = ax$ (see Example 106, p. 55), prove that the orbits are the left cosets of H.

9. H is a subgroup of G. Prove that G acts on G/H if we define, for any $Hx \in G/H$ and any $z \in G$, $(Hx)*z = Hxz$. Show that G/H is transitive (p. 63) and that the stabilizer of Hx is $x^{-1}Hx$. Hence prove that $x^{-1}Hx$ is a subgroup of G, and that if H has finite index, then so has $x^{-1}Hx$, and these indices are equal.

10. Prove that $S(n)$ acts on the set A of all polynomials in n variables $x_1, ..., x_n$ if we define, for $f(x_1, ..., x_n) \in A$ and $\theta \in S(n)$, $f * \theta$ to be the polynomial f^θ given by $f^\theta(x_1, ..., x_n) = f(x_{1\theta}, ..., x_{n\theta})$. Find the orbit of $f = x_1 x_2 + x_3 x_4$ under $S(4)$. Find also the stabilizer H of f and prove that H is a Sylow 2-subgroup of $S(4)$.

11. Prove that $V = \{e, t, u, v\}$ is normal in $A(4)$, that $K = \{e, t\}$ is normal in V, but that K is not normal in $A(4)$.

12. If G acts on A and if H is the stabilizer of an element $a \in A$, show that $H \lhd G$ if and only if H is the stabilizer of every element of the orbit O_a (see Exercise 18, Chapter 5, p. 57).

13. If $H \lhd G$ and $x \in G$ prove that $(Hx)^n = Hx^n$ for every integer n. Show that $(Hx)^n$ is the unit element of G/H if and only if $x^n \in H$.

14. Z is a subgroup of the group $(R, +)$ (Example 81, p. 11), and it is normal because $(R, +)$ is Abelian. Show that an element $x + Z$ $(x \in R)$ has finite order in the group R/Z if and only if x is *rational*, i.e. if $x = m/n$ for some integers m, n.

CHAPTER SEVEN

Homomorphisms

1. Homomorphisms

Our study of groups has been confined until now to things, such as subgroups, quotient sets and quotient groups, which are related to a single given group G. In this chapter we consider how to *compare* two groups G and H. This is done by studying those mappings of G into H, called *homomorphisms*, which 'transform' the group operation of G into that of H.

DEFINITION. Let $(G, \bigcirc)$ and $(H, \times)$ be groups, with group operations $\bigcirc$ and $\times$ respectively. Then a *homomorphism* of G into H is a mapping $\theta : G \to H$ which satisfies the condition

H $$(x \bigcirc y)\theta = (x\theta) \times (y\theta)$$

for all $x, y \in G$.

In other words, a homomorphism is a mapping with the property that it maps the product of x and y to the product of $x\theta$ and $y\theta$, for all x, y in G. Special names are used for special kinds of homomorphisms: a homomorphism $\theta : G \to H$ is called a *monomorphism* if θ is injective, an *epimorphism* if θ is surjective, and an *isomorphism* if θ is bijective. A homomorphism $\theta : G \to G$ of a group into itself is called an *endomorphism*, or an *automorphism* if θ is also bijective; i.e. an automorphism is an isomorphism of G into G, alternatively, it is a permutation of G which is also a homomorphism.

Example 128. The mapping $\phi : R \to R^+$ defined by $x\phi = e^x$ (Example 46, p. 25) is a homomorphism of $(R, +)$ into $(R^+, .)$, for $(x + y)\phi = e^{x+y} = e^x e^y = (x\phi)(y\phi)$, for all $x, y \in R$. Since ϕ is also bijective (Example 48), ϕ is an *isomorphism*.

70

Example 129. Let $(G, .)$ be a group and x an element of G. Define $\theta_x:Z \to G$ by the rule: if $n \in Z$ let $n\theta_x = x^n$. Then θ_x is a homomorphism of $(Z, +)$ into G. For if $m, n \in Z$ we have $(m + n)\theta_x = x^{m+n} = x^m x^n$ (Example 87, p. 46) $= (m\theta_x)(n\theta_x)$.

Example 130. If G is any group, the identity map ι_G is always a homomorphism, so in fact it is an automorphism of G.

Example 131. Let $(G, .)$ be a group and x an element of G. Define $\pi_x:G \to G$ by the rule: if $a \in G$ let $a\pi_x = a^x = x^{-1}ax$. Then π_x is a homomorphism, by the identity $(ab)^x = a^x b^x$ (Example 124, p. 66). Now if $x, y \in G$ we have $(a^x)^y = a^{xy}$ for all $a \in G$ (Example 103, p. 55), i.e. $\pi_x \pi_y = \pi_{xy}$. But also $a^e = a$ for all $a \in G$, which shows that $\pi_e = \iota_G$. Thus for any x, the mapping $\pi_{x^{-1}} = \theta'$, say, satisfies $\pi_x \theta' = \theta' \pi_x = \iota_G$. This proves that π_x is *bijective* (Chapter 3, 6(iii), p. 31), therefore π_x is an automorphism of G; these automorphisms are called *inner automorphisms*. We can prove incidentally, that the set $I(G)$ of all inner automorphisms $\pi_x, x \in G$, is a subgroup of the symmetric group $S(G)$ (Example 79, p. 43). For if $\pi_x, \pi_y \in I(G)$, then $\pi_x \pi_y = \pi_{xy} \in I(G)$ and also $(\pi_x)^{-1} = \theta' = \pi_{x^{-1}} \in I(G)$.

2. Some lemmas on homomorphisms

If G, H are groups both written multiplicatively, then condition **H** for a mapping $\theta:G \to H$ to be a homomorphism reads $(xy)\theta = (x\theta)(y\theta)$, for all $x, y \in G$. Suppose now that $\theta:G \to H$ is a homomorphism.

(i) LEMMA. *If* x, $y \in G$ *then* $(yx^{-1})\theta = (y\theta)(x\theta)^{-1}$.

Proof. Write $u = x\theta$, $v = y\theta$ and $w = (yx^{-1})\theta$. Then by **H**, $wu = v$, so $w = vu^{-1}$, as required. If we take $x = y = e_G$, the unit element of G, then $(y\theta)(x\theta)^{-1} = (e_G\theta)(e_G\theta)^{-1} = e_H$, the unit element of H, and this proves the next lemma.

(ii) LEMMA. $e_G\theta = e_H$, *for any homomorphism* $\theta:G \to H$.

Now put $y = e_G$ in (i). Using (ii), we have now

(iii) LEMMA. $(x^{-1})\theta = (x\theta)^{-1}$, *for any* $x \in G$.

Suppose now that G, H, K are groups.

(iv) LEMMA. *If* $\theta:G \to H$ *and* $\phi:H \to K$ *are both homomorphisms* (*isomorphisms*), *then* $\theta\phi:G \to K$ *is a homomorphism* (*isomorphism*).

Proof. Verify that $\theta\phi$ satisfies **H**, as follows: if $x, y \in G$ then $(xy)(\theta\phi) = ((xy)\theta)\phi = ((x\theta)(y\theta))\phi$ (using **H** for θ) $= ((x\theta)\phi)((y\theta)\phi)$ (using **H** for ϕ) $= (x(\theta\phi))(y(\theta\phi))$. (If θ, ϕ

are also both bijective then $\theta\phi$ is also bijective (Chapter 3, 6(ii), p. 30).)

(v) LEMMA. *If $\theta:G \to H$ is an isomorphism, then $\theta^{-1} \cdot H \to G$ is an isomorphism.*

Proof. We know that θ^{-1} is bijective, so we have only to verify that θ^{-1} is a homomorphism, i.e. that θ^{-1} satisfies **H**. Let u, v by any elements of H. Put $x = u\theta^{-1}$, $y = v\theta^{-1}$, so that $u = x\theta$ and $v = y\theta$. Since θ satisfies **H**, $uv = (x\theta)(y\theta) = (xy)\theta$, i.e. $(uv)\theta^{-1} = xy = (u\theta^{-1})(v\theta^{-1})$, as required.

Example 132. If G, H are both written additively, **H** reads $(x + y)\theta = x\theta + y\theta$, for all $x, y \in G$. Lemma (i) reads $(y - x)\theta = y\theta - x\theta$.

Example 133. If $\theta:G \to H$ is a homomorphism and G, H both written in multiplicative notation, then $(x_1 \ldots x_n)\theta = (x_1\theta) \ldots (x_n\theta)$ for any $x_1, \ldots, x_n \in G$, by applying **H** repeatedly. Taking $x_1, \ldots, x_n$ all equal to x gives $(x^n)\theta = (x\theta)^n$ for any positive integer n; lemmas (ii) and (iii) extend this to any integer n.

Example 134. The set $A(G)$ of all automorphisms of a group G is a subgroup of the symmetric group $S(G)$ (cf. Example 131). For if θ, $\phi \in A(G)$, i.e. if θ,ϕ are isomorphisms of G into G, then by lemmas (iv) and (v) $\theta\phi \in A(G)$ and $\theta^{-1} \in A(G)$. The group $I(G)$ of Example 131 is a subgroup of $A(G)$; we have

$$I(G) \subseteq A(G) \subseteq S(G).$$

3. Isomorphism

DEFINITION. Let G, H be groups. Then we say G and H are *isomorphic* and write $G \cong H$, if there exists an isomorphism $\theta:G \to H$.

Isomorphism is the analogue for groups of similarity for sets (Chapter 3, §7, p. 32). Groups which are isomorphic are considered 'equivalent' from the point of view of group theory; the relation $\cong$ has the properties of an equivalence relation, i.e. (1) $G \cong G$ for any group G, (2) if $G \cong H$ then $H \cong G$, finally (3) $G \cong H$ and $H \cong K$ implies $G \cong K$. These facts are proved, using the lemmas (iv), (v) of §2, just as the corresponding facts for similarity (p. 32).

Groups which are isomorphic are exact 'copies' of each other, as far as their binary operations are concerned. For example the groups $Z/3Z$ and $A(4)/V$ (Examples 126, 127, p. 68) evidently have the 'same' multiplication tables, although they have quite different elements. This 'sameness' comes from the fact that the mapping $\theta:Z/3Z \to A(4)/V$, which takes

$$\bar{0} \to V, \bar{1} \to A, \bar{2} \to P$$

is an *isomorphism*, hence that $Z/3Z \cong A(4)/V$.

A property of groups is said to be *invariant* if whenever G, H are isomorphic groups and G has the property, then also H has it. The *order* of a finite group is invariant. For if $G \cong H$ then certainly $G \simeq H$ (since any isomorphism θ is bijective), and this means that isomorphic groups G, H have the same order. Group theory might be described as the study of properties of groups which are invariant in this sense.

Example 135. Example 128 shows that $(R, +) \cong (R^+, .)$.

Example 136. If $G \cong H$ and G is Abelian, then H is also Abelian. For let u, $v \in H$ and let $\theta:G \to H$ be an isomorphism. If $x = u\theta^{-1}, y = v\theta^{-1}$, then $xy = yx$ since G is Abelian; now apply θ to this equation. By H we get $(x\theta)(y\theta) = (y\theta)(x\theta)$, i.e. $uv = vu$. Thus the property of being Abelian is invariant.

Example 137. The groups $S(3)$ and $Z/6Z$ both have order 6, but are not isomorphic, because $S(3)$ is not Abelian and $Z/6Z$ is.

Example 138. Let $\theta:G \to H$ be any isomorphism, and x, y elements of G, H respectively, such that $x\theta = y$ (hence also $y\theta^{-1} = x$). From Example 133 and lemma 2(ii) we find easily that $x^n = e_G \Leftrightarrow y^n = e_H$, for any integer n. Thus x, y have the same order. We can use this to show that the groups V (Example 127) and $Z/4Z$ are not isomorphic, although both are of order 4 and both are Abelian. For $Z/4Z$ has an element $\bar{1}$ of order 4, and V does not.

4. Kernel and image

Any homomorphism $\theta:G \to H$ gives some connection between the groups G and H. If θ is an isomorphism then this connection is very close, and we know that $G \cong H$. If θ is not an isomorphism we cannot say as much, but the fundamental 'Homomorphism Theorem' below tells us that *any* homomorphism of G into H gives rise to an *isomorphism* of a certain quotient group of G and a certain subgroup of H.

HOMOMORPHISMS

DEFINITION. Let G, H be groups and $\theta:G \to H$ a homomorphism.

The sets (1) $\text{Ker } \theta = \{x \in G | x\theta = e_H\}$,

(2) $\text{Im } \theta = \{x\theta | x \in G\}$

are called the *kernel* and *image* of θ, respectively.

Notice that Ker θ is a subset of G, while Im θ is a subset of H.

THE HOMOMORPHISM THEOREM. *Let* G, H *be groups and* $\theta:G \to H$ *any homomorphism. Then*

(i) Ker θ is a normal subgroup of G.

(ii) Im θ is a subgroup (not necessarily normal) of H.

(iii) If we write $K = Ker\ \theta$, there is an isomorphism

$$\theta^*:G/K \to Im\ \theta$$

defined by the rule: if $x \in G$ then $(Kx)\theta^* = x\theta$. Hence

$$G/Ker\ \theta \cong Im\ \theta.$$

Proof. (i) Let x, $y \in$ Ker θ, so that $x\theta = y\theta = e_H$. Then by §2(i) $(xy^{-1})\theta = e_H e_H^{-1} = e_H$, hence $xy^{-1} \in$ Ker θ, and this shows that Ker θ is a subgroup of G. To prove it is normal in G, use condition N″ (p. 66). If $u \in$ Ker θ and x is any element of G, $(x^{-1}ux)\theta = (x\theta)^{-1}(u\theta)(x\theta) = (x\theta)^{-1}e_H(x\theta) = (x\theta)^{-1}(x\theta) = e_H$, so that $x^{-1}ux \in$ Ker θ, as required.

(ii) Let u, v be any elements of Im θ, we want to prove that $uv^{-1} \in$ Im θ, i.e. that Im θ satisfies condition S (p. 50). By definition of Im θ, there exist $x, y \in G$ such that $x\theta = u$ and $y\theta = v$. By §2(i), $uv^{-1} = (xy^{-1})\theta$, which belongs to Im θ.

(iii) Let x, y be any elements of G. We prove first a lemma.

LEMMA. $Kx = Ky \Leftrightarrow yx^{-1} \in K \Leftrightarrow x\theta = y\theta$.

The first implication is simply the theorem on p. 59, and for the second we have: $yx^{-1} \in K =$ Ker $\theta \Leftrightarrow (yx^{-1})\theta = e_H \Leftrightarrow (y\theta)(x\theta)^{-1} = e_H$ (by §2(i)) $\Leftrightarrow y\theta = x\theta$. This proves the lemma.

The *consistency* of the rule defining θ^* follows from the statement $yx^{-1} \in K \Rightarrow x\theta = y\theta$. From the statement $x\theta = y\theta \Rightarrow Kx = Ky$ we see θ^* *is injective*. From the definition of

Im θ, θ^* *is surjective.* We now show that θ^* *satisfies* **H**: let $x, y \in G$, then $((Kx)(Ky))\theta^* = (Kxy)\theta^* = (Kxy)\theta = (x\theta)(y\theta) = ((Kx)\theta^*)((Ky)\theta^*)$. Therefore θ^* is an isomorphism.

Example 139. If $\theta:G \to H$ is injective, then the only element $x \in G$ such that $x\theta = e_H$ is $x = e_G$ (we know $e_G\theta = e_H$, by §2(ii)), hence Ker $\theta = \{e_G\}$. Conversely, the lemma above shows that if Ker $\theta = \{e_G\}$, then $x\theta = y\theta \Rightarrow x = y$, i.e. θ is injective. Hence a *homomorphism θ is a monomorphism if and only if Ker $\theta = \{e_G\}$*. This is a companion to the theorem that a *homomorphism θ is an epimorphism (i.e. is surjective) if and only if* Im $\theta = H$, which is immediate from the definition of Im θ.

Example 140. If $\{e\}$ is the one-element subgroup of a group G, then $G/\{e\} \cong G$. This is very easy to verify directly, but it is also a consequence of the homomorphism theorem — take $\theta = \iota_G$, and check that Ker $\iota_G = \{e\}$, Im $\iota_G = G$.

Example 141. Classification of cyclic groups. Let H be a cyclic group with unit element e, generated by an element x. Let θ_x be the homomorphism of Z into H defined by $n\theta_x = x^n$ (see Example 129). Then Im $\theta = \{x^n | n \in Z\} = \text{gp}\{x\} = H$, and Ker θ is the set of all integers n such that $x^n = e$. If H is *infinite*, then Ker $\theta = \{0\}$ (see p. 53), hence by the homomorphism theorem $Z/\{0\} \cong H$, i.e. $H \cong Z$. If H is *finite of order m*, then (Example 101, p. 54) Ker θ is the group mZ of all multiples of m, hence $H \cong Z/mZ$. Therefore *every cyclic group is isomorphic to one of*

(1) $Z, Z/Z, Z/2Z, Z/3Z, \ldots$.

No two of these are isomorphic, because they all have different orders. So (1) is a complete list of cyclic groups, up to isomorphism.

Example 142. Alternating groups. Let $x_1, \ldots, x_n$ be n variables ($n \geqslant 2$) and let $\Delta(x_1, \ldots, x_n)$ be the polynomial $\prod_{i<j} (x_i - x_j)$, e.g. $\Delta(x_1, x_2, x_3) = (x_1-x_2)(x_1-x_3)(x_2-x_3)$. Then (see e.g. P. M. Cohn, *Linear Equations*, p. 56, in this series — Δ is called ϕ there), for each permutation θ of $\{1, 2, \ldots, n\}$, the polynomial $\Delta\theta = \Delta(x_1\theta, \ldots, x_n\theta)$ is equal either to Δ or to $-\Delta$; θ is called *even* or *odd*, in these respective cases. Define a mapping $\epsilon:S(n) \to R$ as follows (we shall use 'functional notation' $\epsilon(\theta)$ for the image under ϵ of $\theta \in S(n)$):$\epsilon(\theta) = 1$ if θ is even, $\epsilon(\theta) = -1$ if θ is odd. Thus $\Delta\theta = \epsilon(\theta)\Delta$ for all θ. Then ϵ *is a homomorphism of $S(n)$ into* $(R^*, \cdot)$. For if θ, ϕ are any elements of $S(n)$, $\Delta(\theta\phi) = (\Delta\theta)\phi = \epsilon(\theta)\Delta\phi = \epsilon(\theta)\epsilon(\phi)\Delta$, which shows that $\epsilon(\theta\phi) = \epsilon(\theta)\epsilon(\phi)$, i.e. that ϵ satisfies **H**. Ker ϵ is the set $A(n)$ of all even permutations of $\{1, 2, \ldots, n\}$, and Im ϵ is the two-element set $\{1, -1\}$ (notice that the identity permutation is even, and $\theta = \begin{pmatrix} 1234 \ldots n \\ 2134 \ldots n \end{pmatrix}$ is odd, see P. M. Cohn, *loc. cit.*). The homomorphism theorem now tells us that $A(n) \lhd S(n)$ and that $S(n)/A(n) \cong \{1, -1\}$, and in particular $|S(n)/A(n)| = 2$, i.e. $|A(n)| = \frac{1}{2}|S(n)| = \frac{1}{2}n!$, by Lagrange's theorem. $A(n)$ is called the *alternating group of degree n.* The multiplication table of $A(4)$ is given on p. 80.

HOMOMORPHISMS

Example 143. *Cayley's theorem.* Let x be any element of a group G, and define $\rho_x : G \to G$ by the rule: if $a \in G$ let $a\rho_x = ax$. From the group axioms G1, G2 follow at once (i) $\rho_x \rho_y = \rho_{xy}$ for any x, y in G, and (ii) $\rho_e = \iota_G$. Thus for given x, the mapping $\theta' = \rho_{x^{-1}}$ is inverse to ρ_x, i.e. $\rho_x \theta' = \theta' \rho_x = \iota_G$, hence ρ_x is bijective (Chapter 3, C(iii), p. 31). Hence $\rho_x \in \Gamma(G)$, the group of all permutations on the set G. Define $\gamma : G \to S(G)$ as follows: if $x \in G$ let $x\gamma = \rho_x$. Then (i) shows that γ is a homomorphism. Ker γ is the set of all $x \in G$ such that $\rho_x = \iota_G$, but if $\rho_x = \iota_G$ then $e = e\iota_G = e\rho_x = x$, so Ker $\gamma = \{e\}$. Thus the homomorphism theorem gives $G/\{e\} \cong \operatorname{Im} \gamma$, i.e. $G \cong \operatorname{Im} \gamma$ (see Example 140). This proves Cayley's theorem, that *any group is isomorphic to a subgroup of a symmetric group*, for we know that Im γ is a subgroup of $S(G)$.

5. Suggestions for further reading

On sets: S. Swierczkowski, *Sets and numbers*, in this series.

On groups: W. Ledermann, *Introduction to the theory of finite groups*, Oliver and Boyd, or (more advanced) M. Hall, *The theory of groups*, Macmillan.

An excellent elementary introduction to groups and other algebraic structures is: G. Birkhoff and S. Maclane, *A survey of modern algebra*, Macmillan.

EXERCISES FOR CHAPTER SEVEN

1. If G, H are any groups, define the mapping $\theta : G \to H$ by the rule: $x\theta = e_H$, for all $x \in G$. Prove that θ is a homomorphism.

2. If G is Abelian and n any integer, define $\theta : G \to G$ by the rule: $x\theta = x^n$, for all $x \in G$. Prove that θ is an endomorphism of G.

3. Prove that $I(G) \lhd A(G)$ (Examples 131, 134).

4. Let G be any group of order 4. Prove that either (i) G is cyclic (in which case $G = Z/4Z$ by Example 141), or (ii) $G \cong V$ (p. 80). (Hint: the order of any element x of G must divide 4, Example 117, p. 62. If G has an element of order 4, G is cyclic. So if G is not cyclic, the order of each element $x \neq e$ must be 2.)

5. Prove that $Z \cong mZ$, for any integer $m \neq 0$. Prove that $(R^*, \,.\,)$ and $(R, +)$ are *not* isomorphic (p. 72). (Hint: notice that $(R^*, \,.\,)$ contains an element x of order 2, viz. $x = -1$. If $\theta : R^* \to R$ is an isomorphism, what can $x\theta$ be?)

6. Let N be any normal subgroup of G, and define $\theta : G \to G/N$ by the rule $x\theta = Nx$, for all $x \in G$. Prove that θ is an epimorphism, and that Ker $\theta = N$. (This shows that *every* normal subgroup of G is the kernel of some homomorphism.)

HOMOMORPHISMS

7. Let $G = \text{gp}\{x\}$ be a cyclic group, H any group, and $\theta:G \to H$ any homomorphism. Prove that $\text{Im }\theta$ is cyclic, and $\text{Im }\theta = \text{gp}\{x\theta\}$.

8. With the notation of the last Exercise, prove that for any integer n, $x^n \in \text{Ker }\theta \Leftrightarrow (x\theta)^n = e_H$. Hence show that $\text{Ker }\theta = \text{gp}\{x^m\}$, where $m = 0$ if $\text{Im }\theta$ has infinite order, and $m = |\text{Im }\theta|$ if $\text{Im }\theta$ has finite order. (Use Example 101, p. 54.)

9. Prove that all subgroups of a cyclic group $G = \text{gp}\{x\}$ are cyclic, and that G has at most one subgroup of any given index. (Hint: let N be any subgroup of G. Since G is Abelian, $N \lhd G$. Let $\theta:G \to G/N$ be the homomorphism of Exercise 6. Now use Exercise 8.)

10. For any group G, define $\pi:G \to A(G)$ by the rule: $x\pi = \pi_x$ (see Examples 131, 134) for all $x \in G$. Show that π is a homomorphism, and that $\text{Im }\pi = I(G)$, and $\text{Ker }\pi = \xi(G)$, the set of all elements x of G which commute with every element of G ($\xi(G)$ is called the *centre* of G). Hence show $\xi(G) \lhd G$ and $G/\xi(G) \cong I(G)$.

11. Prove that $\xi(S(n)) = \{\iota\}$ if $n > 2$, and $\xi(S(n)) = S(n)$ for $n = 1,2$.

12. Let $C = \{x_1x_2 + x_3x_4, x_1x_3 + x_2x_4, x_1x_4 + x_2x_3\}$. Show that each $\theta \in S(4)$ induces a permutation $\epsilon(\theta)$ of C, by acting on the elements of C in the manner of Exercise 10, Chapter 6. Prove that the mapping $\epsilon:S(4) \to S(C)$ so defined is a homomorphism, and that $\text{Ker }\epsilon = V = \{e, t, u, v\}$ (p. 80). Now apply the homomorphism theorem (p. 74) to show that $V \lhd S(4)$ and $S(4)/V \cong S(3)$.

Tables*

1. Symmetric group S(3)

ι	α	β	ρ	σ	τ
α	β	ι	σ	τ	ρ
β	ι	α	τ	ρ	σ
ρ	τ	σ	ι	β	α
σ	ρ	τ	α	ι	β
τ	σ	ρ	β	α	ι

S(3) is the group of all permutations of the set $\{1, 2, 3\}$.

$$\iota = \begin{pmatrix} 123 \\ 123 \end{pmatrix}, \; \alpha = \begin{pmatrix} 123 \\ 231 \end{pmatrix}, \; \beta = \begin{pmatrix} 123 \\ 312 \end{pmatrix},$$

$$\rho = \begin{pmatrix} 123 \\ 132 \end{pmatrix}, \; \sigma = \begin{pmatrix} 123 \\ 321 \end{pmatrix}, \; \tau = \begin{pmatrix} 123 \\ 213 \end{pmatrix}.$$

2. Cyclic group of order 6

This is the subgroup of the symmetric group S(6) which is generated by the permutation $x = \begin{pmatrix} 123456 \\ 234561 \end{pmatrix}$. x can be thought of as a clockwise rotation through $\frac{1}{6}$th of a revolution of the 6-spoked 'wheel' of Figure 13.

---

* In these multiplication tables, the unit element is always put in the top left-hand corner; the product $x \bigcirc y$ is the element in row x, column y.

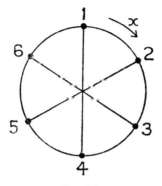

FIG. 13

For any $n \geqslant 0$, x^n rotates the wheel through $\frac{n}{6}$ths revolutions; x^{-n} rotates the wheel anticlockwise by the same amount. $m = 6$ is the least positive integer such that $x^m = e$ (x^6 brings the wheel back to its original position, i.e. x^6 is the identity permutation, which we here denote e), hence x has order 6 and $\text{gp}\{x\} = \{e, x, x^2, x^3, x^4, x^5\}$. Products of these elements can be found using the fact that $x^6 = e$, e.g. $x^3 x^4 = x^7 = x^6 x = x$.

Notice in particular that $x^{-1} = x^{-1} x^6 = x^5$, $x^{-2} = x^4$ etc.

e	x	x^2	x^3	x^4	x^5
x	x^2	x^3	x^4	x^5	e
x^2	x^3	x^4	x^5	e	x
x^3	x^4	x^5	e	x	x^2
x^4	x^5	e	x	x^2	x^3
x^5	e	x	x^2	x^3	x^4

3. Alternating group A(4)

e	t	u	v	a	b	c	d	p	q	r	s
t	e	v	u	b	a	d	c	q	p	s	r
u	v	e	t	c	d	a	b	r	s	p	q
v	u	t	e	d	c	b	a	s	r	q	p
a	c	d	b	p	r	s	q	e	u	v	t
b	d	c	a	q	s	r	p	t	v	u	e
c	a	b	d	r	p	q	s	u	e	t	v
d	b	a	c	s	q	p	r	v	t	e	u
p	s	q	r	e	v	t	u	a	d	b	c
q	r	p	s	t	u	e	v	b	c	a	d
r	q	s	p	u	t	v	e	c	b	d	a
s	p	r	q	v	e	u	t	d	a	c	b

A(4) is the set of all even permutations of S(4) (Example 142, p. 75). The subset $V = \{e, t, u, v\}$ is a subgroup.

V is sometimes called Klein's four-group. In the list of elements below, only the second line of each permutation is given, e.g. (2143) stands for the permutation $\begin{pmatrix} 1234 \\ 2143 \end{pmatrix}$.

$e = (1234)$, $t = (2143)$, $u = (3412)$, $v = (4321)$, $a = (2314)$, $b = ta = (3241)$, $c = ua = (1423)$, $d = va = (4132)$, $p = a^2 = (3124)$, $q = ta^2 = (1342)$, $r = ua^2 = (2431)$, $s = va^2 = (4213)$.

To check the table, first check the products of elements of V. Then any other product can be found from these and the following 'relations' between a and V, which are easily verified: $a^3 = e$, $at = ua$, $au = va$, $av = ta$. For example $bq = tata^2 = t(ua)a^2 = (tu)a^3 = ve = v$.

List of Notations

$x \in A$ x belongs to (is an element of) the set A, p. 1

$|A|$ order (number of elements) of finite set A, p. 2

$B \subseteq A$ B is a subset of A, p. 3

$A \cap B, B \cup A$ intersection, union of sets A, B, pp. 4, 6

$\cap \mathscr{S}, \cup \mathscr{S}$ intersection, union of a set $\mathscr{S}$ of sets, p. 11

$A - B$ difference of sets A, B, p. 8

$A \times B$ product of A, B, p. 10

$\mathscr{B}(X)$ set of all subsets of X, p. 11

$\varnothing$ the empty set, p. 5

$\theta : A \to B$ θ is a mapping of A into B, p. 24

$x\theta$ image of x under the mapping θ, p. 23

θ^{-1} inverse of the bijection θ, p. 27

ι_A identity map on set A, p. 29

$A \simeq B$ A is similar to B, p. 32

$S(A)$ set (or group) of all permutations of A, pp. 31, 43

$S(n)$ set (or group) of all permutations of $\{1, ..., n\}$, pp. 32, 43

gp X subgroup generated by X, p. 52

$G/H, G \backslash H$ set of right, left cosets of H in G, p. 58, or quotient group, p. 67

$H \triangleleft G$ H is a normal subgroup of G, p. 65

$G \cong H$ G is isomorphic to H, p. 72

Ker θ, Im θ kernel, image of the homomorphism, θ, p. 74

Z set of all integers, p. 2, or the group $(Z, +)$, p. 44

R set of all real numbers, p. 2, or the group $(R, +)$ p. 44

R^{+} set of positive real numbers, p. 3

R^{*} set of all non-zero real numbers, or the group $(R^{*}, .)$, p. 44

Answers to Exercises

Chapter 1. 10. No. 12. Q_1.

Chapter 2. 1. (i) Eq.; (ii) Sym.; (iii) Ref., sym.; (iv) Ref., trans. 6. 1. 8. $A \cap B$ is the congruence class of 7 mod 12.

Chapter 3. 1. 6 are injective. $s!/(s-r)!$. 2. (i) neither; (ii) $(a),(b)$; (iii) (b); (iv) (a). 3. $y\theta^{-1} = \sqrt{y}$, No. 4. $y\theta^{-1} = a^{-1}(y-b)$. 5. They commute if $bc + d = ad + b$. 6. No. 8. No.

Chapter 4. 1. (i) 8; (ii) 8; (iii) 4; (iv) 4; (6 are ass. and comm., all with units or zeros are ass., 2 have all four properties). 2. Neither; 7. Not unless $|A| = 1$.

Chapter 5. 1. Yes. 2. One way is to arrange that $1 \bigcirc 1 = 2, 2 \bigcirc 2 = 3, ..., n \bigcirc n = 1$ (other products arbitrary). 4. $H^* = \{x/y \,|\, x,y$ non-zero integers$\}$. 13. Order of e is 1, of t, u, v is 2, others have order 3. 14. 1, -1. $x, x^3, x^5\, x^7$. 16. For $\xi = \iota$, $C(\xi) = S(3)$; for $\xi = \alpha$ or β, $C(\xi) = \{\iota, \alpha, \beta\}$, for $\xi = \rho$, σ or τ, $C(\xi) = \{\iota, \xi\}$. 17. Let $H = \text{gp}\,\{\beta\} = \{\iota, \beta, \beta^2, \beta^3\}$, and $\theta = \begin{pmatrix} 1234 \\ 1432 \end{pmatrix}$, Then $N\{\beta, \beta^{-1}\} = H \cup H\theta$ (order 8).

Chapter 6. Right: H, Ht, Hu, Hv. Left: H, tH, uH, vH. 2. The one-element subsets of G. Only G itself. 7. $\{e\}$, $\{t, u, v\}$, $\{a, b, c, d\}$, $\{p, q, r, s\}$. 10. O_f is the set C of Exercise 12, p. 77. $H = V \cup Vy$, where $V = \{e, t, u, v\}$ and $y = \begin{pmatrix} 1234 \\ 2134 \end{pmatrix}$ (H has order $8 = 2^3$, which is the highest power of 2 dividing $24 = |S(4)|$).

Index

INDEX